FEASIBILITY OF VEHICLE-BASED SENSORS TO DETECT DROWSY DRIVING AND ALCOHOL IMPAIRMENT

(WITH ACCOMPANYING CD-ROM)

ALCOHOL AND DRUG ABUSE

Additional books in this series can be found on Nova's website
under the Series tab.

Additional e-books in this series can be found on Nova's website
under the e-book tab.

FEASIBILITY OF VEHICLE-BASED SENSORS TO DETECT DROWSY DRIVING AND ALCOHOL IMPAIRMENT

(WITH ACCOMPANYING CD-ROM)

MEAGHAN SADLER
EDITOR

nova
publishers
New York

Library of Congress Cataloging-in-Publication Data

ISBN: 978-1-61728-349-9

Published by Nova Science Publishers, Inc. † New York

CONTENTS

PREFACE

Drowsy driving is a significant contributor to death and injury crashes on our Nation's highways, accounting for more than 80,000 crashes and 850 fatalities per year. The successful detection of drowsiness is a crucial step in implementing mitigation strategies to reduce the cost to society of drowsy driving. Despite persistent efforts at the local, state, and federal levels, alcohol-impaired driving crashes also contribute to approximately 31% of all traffic fatalities. Although regulatory and educational approaches have helped reduce alcohol-related fatalities, other approaches merit investigation. This book discusses the feasibility of vehicle-based sensors to detect both drowsy driving and alcohol-impaired driving.

Chapter 1 –Building upon prior research in detecting impairment from alcohol and distraction, the goal of this research was to determine the extent to which alcohol impairment algorithms could detect drowsiness and distinguish it from alcohol impairment. Data were collected from seventy-two participants during daytime (9 a.m. - 1 p.m.), early night (10 p.m. – 2 a.m.), and late night (2 a.m. - 6 a.m.) sessions to provide data for algorithm testing and refinement. Driving data indicated a complex relationship between driving performance and conditions associated with drowsiness: compared to daytime session, driving performance improved during the early night session, before degrading during the late night session. This non- linear relationship between continuous time awake, subjective assessments of drowsiness and driving performance has the potential to complicate the early detection of drowsiness. Drowsiness, as indicated by unintended lane departures, occurred in all sessions and demonstrated a transient nature. Algorithms based on lane position and steering wheel data, which can be obtained inexpensively, were best at predicting drowsiness related lane departures. Alcohol detection algorithms were not successful in detecting drowsiness but could be retrained to do so. Rather than one algorithm being generalized to detect multiple impairments, these results indicate that specialized algorithms might co- exist and allow one to detect and differentiate alcohol and drowsy-impaired driving. These findings provide a better understanding of the relationship between impairment from alcohol and drowsiness and lay the foundation for detecting and differentiating among impairment from alcohol, drowsiness, fatigue and drugs.

Chapter 2 – Despite persistent efforts at the local, state, and federal levels, alcohol-impaired driving crashes still account for 31% of all traffic fatalities. The proportion of fatally injured drivers with blood alcohol concentrations (BAC) greater than or equal to 0.08% has remained at 31-32% for the past ten years. Vehicle-based countermeasures have the potential to address this problem and save thousands of lives each year. Many of these vehicle-based countermeasures depend on developing an algorithm that uses driver performance to assess impairment. The National Advanced Driving Simulator (NADS) was used to collect data needed to develop an algorithm for detecting alcohol impairment. Data collection involved 108 drivers from three age groups (21-34, 38-51, and 55-68 years of age) driving on three types of roadways (urban, freeway, and rural) at three levels of alcohol concentration (0.00%, 0.05%, and 0.10% BAC). The scenarios used for this data collection were selected so that they were both representative of alcohol-impaired driving and sensitive to alcohol impairment. The data from these scenarios supported the development of three algorithms. One algorithm used logistic regression and standard speed and lane-keeping measures; a second used decision trees and a broad range of driving metrics that are grounded in cues NHTSA has suggested police officers use to identify alcohol-impaired drivers; a third used a support vector machines. The results demonstrate the feasibility of a vehicle-based system to detect alcohol impairment based on driver behavior. The algorithms differentiate between drivers with BAC levels at and above and below 0.08%BAC with an accuracy of 73 to 86%, comparable to the standardized field sobriety test. This accuracy can be achieved with approximately eight minutes of driving performance data. Differences between drivers and between roadway situations have a large influence on algorithm performance, which suggests the algorithms should be tailored to drivers and to road situations.

In: Feasibility of Vehicle-Based Sensors
Editor: Meaghan Sadler

ISBN: 978-1-61728-349-9
© 2014 Nova Science Publishers, Inc.

Chapter 1

ASSESSING THE FEASIBILITY OF VEHICLE-BASED SENSORS TO DETECT DROWSY DRIVING[*]

Timothy Brown, John Lee, Chris Schwarz, Dary Fiorentino and Anthony McDonald

ABSTRACT

Drowsy driving is a significant contributor to death and injury crashes on our Nation's highways, accounting for more than 80,000 crashes and 850 fatalities per year. The successful detection of drowsiness is a crucial step in implementing mitigation strategies to reduce the cost to society of drowsy driving. Building upon prior research in detecting impairment from alcohol and distraction, the goal of this research was to determine the extent to which alcohol impairment algorithms could detect drowsiness and distinguish it from alcohol impairment. Data were collected from seventy-two participants during daytime (9 a.m. - 1 p.m.), early night (10 p.m. – 2 a.m.), and late night (2 a.m. - 6 a.m.) sessions to provide data for algorithm testing and refinement. Driving data indicated a complex relationship between driving performance and conditions associated with drowsiness: compared to daytime session, driving performance improved during the early night session, before degrading during the late night session. This non-linear relationship between continuous time awake, subjective assessments of drowsiness and driving performance has the potential to complicate the early detection of drowsiness. Drowsiness, as indicated by unintended lane departures, occurred in all sessions and demonstrated a transient nature. Algorithms based on lane position and steering wheel data, which can be obtained inexpensively, were best at predicting drowsiness related lane departures. Alcohol detection algorithms were not successful in detecting drowsiness but could be retrained to do so. Rather than one algorithm being generalized to detect multiple impairments, these results indicate that specialized algorithms might co-exist and allow one to detect and differentiate alcohol and drowsy-impaired driving. These findings provide a better understanding of the relationship between impairment from alcohol and drowsiness and lay the foundation for detecting and differentiating among impairment from alcohol, drowsiness, fatigue and drugs.

---

[*] This is an edited, reformatted and augmented version of a National Highway Traffic Safety Administration sponsored document, DOT HS 811 886, issued February 2014.

EXECUTIVE SUMMARY

The most notable findings from this study include:

- Algorithms based on driving performance measures could detect impairment due to drowsiness.
- Algorithms designed to detect alcohol-impaired driving were not well suited to detecting drowsiness-impaired driving.
- The time scale for detection of impairment from drowsiness must be shorter than for alcohol impairment, due to the transient nature of drowsiness.
- Performance-based detection algorithms have the potential to outperform more traditional methods such as percentage of eye closure (PERCLOS), at a lower cost.

Background

Drowsy driving is a significant contributor to death and injury crashes on our Nation's highways accounting for more than 80,000 crashes and 850 fatalities per year. Recent research using data from the 100-car naturalistic study found that drowsy driving contributed to 22 percent to 24 percent of crashes and near-crashes observed. According to the National Sleep Foundation's 2009 annual Sleep in America survey, 28 percent of drivers had driven drowsy at least once per month in the past year. Of those who drove while drowsy, 28 percent have fallen asleep. The rate of drowsy driving and the severity of the resultant crashes give clear cause for concern and research continues to be needed to help reduce the numbers of lives lost due to drowsy driving. Previous research in detecting alcohol impairment showed that algorithms based on driving performance metrics could reliably tell the difference between an impaired driver from an unimpaired driver based on a signature pattern of lane position and steering. Algorithms such as these could be implemented as vehicle-based safety systems to detect impairment from drowsiness.

Objectives

This report describes efforts completed in Phase 1 of the Driver Monitoring of Inattention and Impairment Using Vehicle Equipment (DrIIVE) program to develop and assess algorithms for the detection of drowsy driving. It begins with the application of alcohol detection algorithms to the drowsiness impairment. Specific objectives include:

- Evaluate previously developed algorithms designed to detect alcohol impairment for their ability to detect drowsiness.
- Determine if algorithms designed to detect alcohol impairment can be generalized to detect both alcohol and drowsiness.
- Determine if algorithms can distinguish between impairment caused by alcohol and drowsiness.

- Determine if real-time algorithms can reliably detect drowsiness in advance of a drowsiness-related mishap, and do so better than event-based algorithms.

Method

Data were collected from 72 participants in the National Advanced Driving Simulator on three drives over two visits: one daytime drive between 9 a.m. and 1 p.m.; two nighttime drives with an early night drive between 10 p.m. and 2 a.m. and a late night drive between 2 a.m. and 6 a.m. Drivers were divided into equal groups by age (21 to 34, 38 to 51, and 55 to 68) and gender. The participants drove a scenario representative of a nighttime drive home from an urban area for a total drive time of approximately 35 minutes. The drives started with an urban segment composed of a two-lane roadway through a city with posted speed limits of 25 to 45 mph with signal-controlled and uncontrolled intersections. A suburban segment followed that consisted of a four-lane divided expressway with a posted speed limit of 70 mph. The drives continued with a rural segment composed of a two-lane undivided road with curves, ending with a ten-minute long drive on a section of straight rural roadway. Drivers' control inputs, vehicle state, driving context, and driver state were captured in representative driving situations, with precise control and in great detail.

Results

The objectives were addressed with two broad sets of analyses. The first focused on whether drowsiness affected performance. The second focused on detection of impairment. These analyses show the simulator and scenario to be sensitive to drowsiness, and that algorithms can detect drowsiness-related impairment.

Driving data indicated that a complex relationship exists wherein driving performance improves with low levels of drowsiness in the early night session before degrading in the late night session. This non-linear relationship between continuous time awake, subjective assessments of drowsiness and driving performance has the potential to complicate the early detection of drowsiness. Drowsiness, as indicated by unintended lane departures, occurred in all conditions and highlights the transient nature of the impairment from drowsiness. Alcohol detection algorithms were not successful in detecting drowsiness but could be retrained to do so. Rather than one algorithm generalized to detect multiple impairments, these results indicate that specialized algorithms might co-exist and allow one to detect and differentiate alcohol and drowsy-impaired driving.

Recommendations and Conclusions

This study demonstrates the feasibility of detecting drowsiness with vehicle-based sensors. Results show that the differences in the manifestation of alcohol and drowsiness impairment do not allow for a single algorithm to detect both types of impairment; however similar algorithms trained independently may be successful. To detect impairment due to either alcohol or drowsiness, a more complex approach is necessary where separate algorithms are combined to work with each other. These results suggest promise in a vehicle-based approach to impairment detection including multiple types of impairment.

Future research should focus on examining distraction related impairment to evaluate the extent to which distraction can be detected when drivers are impaired from alcohol or drowsiness, and the extent to which impairment from alcohol, drowsiness and distraction can be distinguished. Then other types of impairments may also be considered, such as drugs and age-related cognitive decline.

Additional research should evaluate the extent to which existing impairment detection algorithms are capable of detecting impairment from medications or illicit drugs. Many over the counter medications are known to produce drowsiness; however, because these medications produce a more uniform level of drowsiness compared to the transient nature of the natural onset of drowsiness, this type of impairment should be tested to determine if the algorithms developed to detect drowsiness as part of this research would detect driving impaired by medications or illicit drugs.

1. BACKGROUND

Exact counts of the number of crashes caused by drowsiness are hard to obtain due to the use of varying methodologies. The Gallup organization surveyed drivers and estimated that during the 5 years prior to 2002 as many as 1.35 million drivers may have been involved in drowsy-driving- related crashes (Royal, 2003). A National Highway Traffic Safety Administration report of crash report data from 2005 to 2009 attributed 83,000 crashes per year and 886 fatal crashes per year to drowsy, fatigued, or sleeping drivers. Over the 5-year period these causes resulted in 5,021 fatalities. Similar variability in research methods, driver populations, and findings is seen for the percentage of drowsy driving crashes. The 100-car naturalistic driving study found that drowsy driving contributed to 22 percent to 24 percent of crashes and near-crashes observed (Klauer et al., 2006). In a report to Congress, NHTSA stated that 3.2 percent of crashes were related to actual sleep (NHTSA, 2008). An estimated 1 percent of all large-truck crashes, 3 to 6 percent of fatal heavy-truck crashes, and 15 to 33 percent of fatal-to-the-truck-occupant-only crashes have been attributed to driver fatigue as a primary factor (Knipling & Shelton, 1999). Although the methodologies result in different estimates, all point to a significant problem.

According to the National Sleep Foundation's 2009 annual Sleep in America survey, 28 percent of drivers had driven drowsy at least once per month in the past year. Of those that drove while drowsy, 28 percent have fallen asleep (NSF, 2009). A survey conducted in 2003 found that 37 percent of drivers have nodded off for at least a moment or fallen asleep while driving at least once in their driving careers, while 8 percent of them had done it in the last 6 months. Of those encountering an episode of nodding off, 58 percent of drivers were on a multilane interstate highway, and 92 percent of them were startled awake and of those who were startled awake, 33 percent wandered into another lane or shoulder, 19 percent crossed the centerline and 10 percent ran off road (Royal, 2003). Drowsy driving is not only common in the United States, it was found that one in five Canadian drivers have admitted to nodding off or falling asleep at least once while driving (Beirness, 2005) and that driver fatigue contributes to at least 9 to 10 percent of crashes in the United Kingdom (Maycock, 1997).

Clearly, there is cause for concern about the rate of drowsy driving and the resultant crashes, injuries and fatalities. Research continues to be needed to develop technological approaches that will help reduce the numbers of lives lost due to drowsy driving. The present aim is to extend Impairment Monitoring to Promote Avoidance of Crashes using Technology or IMPACT, a program of research into detecting alcohol-impaired driving based primarily upon vehicle-based measures to the domain of drowsy driving (Lee et al., 2010). IMPACT has developed alcohol detection algorithms for all drivers (general algorithms) and algorithms that take into account individual driving differences (individualized algorithms). This work explores how well the previously developed algorithms that detect impairment from alcohol are able to detect drowsiness, and how to best to modify those algorithms, if necessary, to detect both. The algorithms that were previously developed to detect alcohol impairment were effective at levels comparable to the Standardized Field Sobriety Test in 8 to 25 minutes. One algorithm used logistic regression of standard speed and lane-keeping measures; a second used decision trees and a broad range of driving metrics that were grounded in cues NHTSA has suggested police officers use to identify alcohol-impaired drivers; a third used support vector machines and the standard deviation of lane-keeping.

To better place these algorithms in the context of existing research, four research questions must first be addressed:

- Can algorithms designed to detect alcohol impairment and distraction also detect drowsiness?
- Can algorithms designed to detect alcohol impairment be generalized to detect both alcohol and drowsiness?
- Can algorithms distinguish between alcohol and drowsiness-related impairment?
- Do real-time algorithms perform better in detecting drowsiness in advance of a drowsiness-related mishap?

The following sections describe what has been learned from previous research that can help to inform this project.

Terminology

While this project focuses on studying drowsy drivers as opposed to fatigued drivers, it should be noted that while reviewing the literature, the words fatigue and drowsiness were often used interchangeably. For example, the recent NHTSA Traffic Safety Facts on Drowsy Driving defined a drowsy driving crash as one "in which the driver was reported as drowsy, sleepy, asleep, or fatigued" (NHTSA, 2011). For the purpose of this research, drowsy is defined as instances where the driver wishes to sleep, and fatigued as instances where the driver wishes to cease working (driving). In reviewing the open literature, while an author may have used the term fatigued, the keywords of the publication generally included drowsiness or related physiological and cognitive indices of drowsiness, such as attentional resources, vigilance, or effort. This was also true in the reverse, as authors that used the term drowsiness had key words that included fatigue, inattention, fatigued driving, and sustained attention. Fatigue and drowsiness can co-occur. However, in the following review of literature, careful attention was paid to ensure that when articles concerning fatigue were reviewed, the fatigue symptoms and methodology were indicative of a study of drowsy driving. All studies solely of physical fatigue were excluded from the review. Exclusion of all articles that used the term fatigue, however, would have produced a review that does not yield a full understanding of the behavioral indicators of drowsy driving and the environments in which those indicators are found. For the purposes of this review discussion, fatigue can be interpreted as synonymous with drowsy driving.

1.1. Scenario Characteristics

The difficulty of different driving scenarios or situations may depend upon whether a driver is impaired, and if so, the type of impairment. Alcohol impairment is generally the most understood due to the precision of its measurement (breath or blood alcohol concentration), specific legal limit, and its consequent use as a comparison for other types of impairment research. However, different types of impairment manifest in different ways, and just because a driver may find a scenario challenging when impaired with alcohol, does not necessarily mean that a drowsy driver will find it challenging. This section describes why certain scenarios may be more challenging to drowsy drivers than others. The characteristics of such scenarios that are difficult for drowsy drivers can be categorized as ones that affect either endogenous (internal) or exogenous (external) contributors to drowsiness. Circadian variation, time on task, and lack of sleep are considered endogenous whereas scenario characteristics represent exogenous factors (Thiffault & Bergeron, 2003). These authors demonstrated that unpredictable roadside scenery can disrupt the deleterious effects of an otherwise monotonous driving environment. Their findings suggest that "monotony may exacerbate the impact of late night driving, whilst overloaded roadside environments may generate arousal levels that counteract this effect" (p. 382). Similarly, straight road conditions are more challenging to drowsy drivers than curved roads (Matthews & Desmond, 2002).

Overall, these studies suggest that the most challenging driving situation for a drowsy driver would be a long, low demand, predictable driving environment with little driver intervention required. A scenario with a long rural straightaway, little interaction with other traffic and no curves would be consistent with the evidence presented. Additionally, this would suggest that roads with few changes in the surrounding roadway environment such as buildings and signage would also prove more challenging to a drowsy driver. Such situations that come towards the end of a drive are likely to place a greater demand on a drowsy driver because drowsiness tends to increase as time on task increases.

1.2. Reliable and Sensitive Vehicle-Based Indicators

Although there are many measures of driver fatigue and drowsiness, those that are commonly studied are generally perceptual, biological, physiological, or performance based. Vehicle-based indicators of drowsy driving have been less prevalent among studies assessing driver drowsiness or fatigue, and their associated effects on performance. However, simple functions of driving performance such as steering wheel movements, lateral shifts, standard deviation of lane position, and frequency of line crossings and have all been used to measure the effects of drowsiness on driving performance

A review article by Liu, Hosking, and Lenne (2009) summarizes the effects on driving performance measures of driver drowsiness or fatigue based on 17 studies published in peer-reviewed journals in which at least one objective vehicle-based measure was reported. Overall, the reviewed literature indicated an increase in lane departures with increased drowsiness. Moreover, the average standard deviation of lane position (SDLP) and mean absolute value of steering wheel angle and standard deviation of steering wheel movements were shown to increase with drowsiness. It was noted that the current body of knowledge also associates drowsiness with increases in standard deviation in speed and variation in speed from the speed limit, but not consistently. The authors also point out that the research does not present analyses of time histories as the basis of determining drowsiness, but instead focuses on overall averages across entire test periods. This research provides a foundation for focusing the review of indicators of drowsy driving.

Steering wheel movements and the resultant heading error have shown to be reliable indicators of drowsiness. A review of literature related to fatigued and drowsy driving by Barr et al. (2003) found changes in steering behavior are associated with a "driver's state of impairment." Platt (1963) and Safford and Rockwell (1967) found that reduced driver capabilities were associated with an increase in steering reversal rates. Matthews and Desmond (2002) categorized steering reversals into three levels; fine (<2 degrees), medium (2-10 degrees), and coarse (>10 degrees). This is similar to the categories defined by Wilson and Greensmith (1983) that defined fine steering reversals as those less than 2 degrees and course steering reversals as those greater than 12 degrees. It was assumed that coarse reversals reflect reactive responses to lateral drift while fine, and even medium reversals reflect controlled activity (Matthews & Desmond 2002; Mackie & Miller, 1978). One of the most prevalent measures of drowsy driving throughout the literature is SDLP. Liu et al. (2009) point out that there are variations of this measure that index different aspects of driver performance. Precision is defined as the ability of the driver to maintain straight driving, independent of their location within the lane or with respect to the center of the lane. On the other hand, bias is defined as the driver's ability to accurately track the center of the lane. While both of these variations are used in the literature as measures of standard deviation of lane position, it is recommended that they be reported as separate measures (Liu et al., 2009). For the purposes of this report, SDLP will be defined as the deviation from the center of the lane unless otherwise noted.

Many researchers have shown that SDLP increases with increased drowsiness. Arnedt et al. (2001) showed that hours of wakefulness are predictive of changes in SDLP. Their research found that 19 and 22 hours of wakefulness resulted in SDLPs that were consistent with impaired performance at .05 grams per deciliter and .08 g/dL blood alcohol concentration (BAC), respectively. Using a time on task approach and partial sleep deprivation, Otmani et al. (2005) found that SDLP was greater with partial sleep deprivation than with normal sleep, and that it increased over the course of a 90-minute drive. The partial sleep deprivation condition that used moderate sleep restriction during the night prior to the driving session consisted of approximately 12 hours of wakefulness in the 16-hour period before driving. Subjects were allowed to sleep only from 3 to 7 a.m. with driving occurring during the "post-lunch dip period between 2 and 4 p.m." Another type of study examining the effects of caffeine by De Valck and Cluydts (2001) showed that SDLP was sensitive to both the effects of hours of sleep and caffeine: increased SDLP with less sleep, and decreased SDLP after using caffeine. It should be noted, however, that SDLP is also affected by substances such as alcohol and distraction as documented in the IMPACT program (Lee et al., 2011a), and the Distraction Detection and Mitigation Through Driver Feedback (Lee et al., 2011b) final reports. While this metric may facilitate multiple impairment detection, it may not be very useful for distinguishing among them.

Inappropriate line crossings (lane departures) also increase with drowsy driving. Philip et al. (2005) found that the number of inappropriate line crossings, defined as crossing one of the lateral highway lane markers, increased for sleep-deprived drivers as opposed to well-rested drivers. Speed control is another measure where research has shown differences. This measure has not been reported as often as have lateral control measures; however, a number of researchers have found it to be sensitive to the effects of drowsiness. Arndt (2001) also found that speed variability increased with hours of wakefulness. Specifically, he found greater variability after 20 hours of wakefulness than after 16 hours; however, when comparing the effect of alcohol, the effect of hours of wakefulness is less than the effect of alcohol at the .08 g/dL BAC. De Valck and Cluydts (2001) showed that deviation from the speed limit increased with less sleep, but decreased when using caffeine under these conditions.

Overall, it appears that there are potentially several diagnostic vehicle-based indicators of drowsiness with lateral control measures the most promising. Across the studies reviewed by Liu et al., the most sensitive and reliable indicator appears to be lateral vehicle control, specifically SDLP.

1.3. Current Algorithms

This project builds from the detection of impairment due to alcohol intoxication, and compares the performance for alcohol detection and drowsiness detection algorithms to correctly identify episodes of drowsy driving based upon a protocol of prolonged wakefulness. First, consider the methods currently proposed for detecting alcohol impairment. A review of the literature indicates that the primary focus of algorithm development to detect alcohol impairment has been on interlock systems. This includes approaches such as the currently deployed breath-based alcohol detection, and newer technologies such as sniffers to detect the presence of breath- alcohol from the driver (Nissan, 2011), transdermal ethanol detection (Webster, 2007) and tissue spectrometry (Ridder et al., 2008). Lee et al. (2010) demonstrated three algorithms that use vehicle control measures such as variability in lane position and speed to predict impairment from alcohol above the legal limit. These algorithms were implemented to detect impairment from alcohol by considering driving performance over a period of similar driving demand (event). Performance metrics primarily included lane keeping and speed control, which were combined to predict impairment.

Several contrasts can be observed between algorithms that are sensitive to alcohol impaired and drowsy driving. Whereas algorithms to detect alcohol have been validated by directly measuring BAC, there is no corollary measure of drowsiness. Instead, drowsiness research has primarily focused on eye behavior such as PERCLOS, or brain activity (Dinges, Mallis, Maislin, & Powell, 1998). When considering driving data that could indicate impairment, the alcohol detection algorithms focused on changes in variability of lane keeping and speed control. However, research indicates that the safety degradations associated with drowsiness may not be present at lower levels of drowsiness (Fairclough & Graham, 1999). In this study, while near lane crossings were more common for drivers drowsy from partial sleep deprivation (only 4 hours of sleep the preceding night), those with full sleep deprivation (no sleep the preceding night) had more frequent actual lane crossings. Both groups of drowsy drivers had a lower steering wheel reversal rate than did control drivers or drivers under the influence of alcohol. In general, this suggests a need to look beyond events directly relevant to safety to detect drowsiness (Fairclough & Graham, 1999). This conclusion is born out of the approaches used in several drowsy driver detection algorithms that focus not only on vehicle performance measures, but also on driver input measures. (Tijerina et al., 1999; Mattsson, 2007).

As the goal of this literature review is to inform the choice of algorithms for comparison to algorithms from Lee et al. (2010), the following sections focus on presenting typical examples of the various approaches that have been attempted. For the purposes of this review, approaches are described in terms of a broad grouping of algorithms that seek to identify similar signatures of drowsiness. The approaches discussed in this review include driver-based, vehicle-based, and combination algorithms.

When algorithm *accuracy* is reported, it is defined as the total correct classifications (hits and correct rejections) relative to all classifications (hits, misses, false alarms and correct rejections). *Specificity* is defined as the ratio of correct rejections to the total number of instances where no drowsiness was present (false alarms + correct rejections). *Sensitivity* is defined as the ratio of hits to the total number of instances where drowsiness was present (misses + hits). The following sections relate the algorithms compared in this study to those found in the literature. Additional details on the algorithms can be found in Appendix A.

1.3.1. Driver-Based Algorithms

The first approach to detecting drowsy driving focused on observing ocular measures of driver drowsiness rather than its manifestation in driving performance. In 1998, NHTSA published an evaluation of several approaches for detecting drowsy drivers based on monitoring the driver (Dinges, Mallis, Maislin, & Powell, 1998). The authors identify these systems as "operator- centered, in-vehicle, [and] fatigue-monitoring technologies (p. 16)," which seek to measure behavioral manifestations of drowsiness. This study examined several different approaches comparing algorithm predictions to performance lapses. It found that PERCLOS was the most reliable indicator of drowsiness in terms of consistent classification. Head position, blinks, and electroencephalograms (EEGs) were found to be less generally applicable across drivers. This effectiveness is likely associated with its general construct validity: measuring when the driver's eyes are closed is a very effective way of identifying when drivers are falling asleep. While reliable, it may provide identification too late to prevent a crash. Additionally, the authors suggest that to improve successful identification of drowsy drivers, a combination of two generally well performing algorithms that complement each other may work best. This approach helps deal with issues associated with a particular algorithm having difficulty with a particular individual. The redundancy of a second algorithm provides a method for detecting drowsiness when individual differences prevent the primary algorithm from working well. This approach was developed in IMPACT (Lee et al. 2010) for alcohol impairment, but in the evaluation, the primary algorithms succeeded often, preventing evaluation of the secondary algorithms.

With increasing video processing capabilities, new approaches to identifying driver drowsiness have emerged that can take into account more complex facial information. Ji et al. (2004) propose an approach that uses a variety of facial information including: head pose, gaze movement, PERCLOS, and facial expression to provide an estimate of level of fatigue. This approach is reliant on being able to extract the information from the video of the driver, and systematically combine the information to predict drowsiness. The facial expression method used is a "feature-based facial-expression-analysis algorithm," that focuses on the driver's eyes and mouth. They report that current work focuses on detecting yawning. Overall, the authors successfully detected drowsiness by comparing a composite measure of fatigue with response time across a variety of drivers of different ages, genders and ethnicities. They report robust, reliable and accurate results; however, specific details concerning their algorithm's performance across individual drivers, and specific metrics such as sensitivity and specificity were not provided in the paper. Thus, it is difficult to gauge the effectiveness of their particular approach.

1.3.2. Vehicle-Based Algorithms

Evaluations of vehicle-based performance measures have shown varying degrees of success. Based upon the findings described about the sensitive indicators of drowsiness above, it is not surprising that many of the efforts to predict impairment focus on lateral control.

Wierwille et al. (1996) proposed a vehicle-based approach to estimate PERCLOS (ePERCLOS) through a combination of measures of steering wheel activity, lane position, and lateral velocity over a three-minute window. This study builds upon the prior successful use of PERCLOS to predict decrements in performance associated with drowsiness (Wierwille et al., 1994). The advantage of this approach is that it does not necessitate the verification of drowsiness; however, this is gained at the risk of misclassifying, if the PERCLOS algorithm fails to accurately capture the actual state of the driver. Using this approach, Wierwille reported a classification accuracy of 96 percent in a simulator study. Tijerina et al. (1999) evaluated this algorithm's reliability in a study with 8 drivers on the road. They found similar results with a reported classification accuracy of 89 percent, indicating that the simulator research transferred well to on-road prediction of PERCLOS.

Tijerina et al. (1999) also evaluated options for improving the performance of a modified, ePERCLOS algorithm. Their approach, BEST ePERC, uses only lane exceedances or excursions (proportion of time out of lane) and variance in lane position to predict PERCLOS and drowsiness. This approach resulted in fewer false alarms, but also fewer true positives than the original.

In a master's thesis, Mattsson (2007) examined the ability of lane position measures to accurately predict drowsiness. A variety of measures of lane position were evaluated and included in a multi equation algorithm with the algorithm selected based upon the data available. The author evaluated the algorithm's performance against drivers' self-reported drowsiness on the Karolinska Sleepiness Scale (KSS). The algorithm was designed to predict KSS values greater than 8 (8 or 9), and proved most accurate when predicting either a reported sleepiness of 8 or 9 on the nine point scale.

Another approach focused on steering wheel behavior to predict when a driver was drowsy. King et al. (1998) described three types of functions that were used to develop the fatigue prediction: time-based, frequency-based, and phase-based. For example, one time-based measure, amplitude duration squared theta, uses the durations found between pairs of consecutive crossings of zero steering wheel angle (i.e., steering reversals). Two phase-based predictors were based on the relationship of the steering wheel angle to its velocity. These predictors were the most successful at detecting periods of fatigue, which was identified through video review on straight road segments of those evaluated. This algorithm has not been extended to work on curves or turns.

1.3.3. Combination Algorithms

More recently, efforts have been made to combine driver-based and vehicle-based performance measures in algorithms that predict drowsiness. One approach that is currently under development is PERCLOS+. This algorithm merges PERCLOS over a 3-minute window with lane deviations over a 1-minute window (Hanowski, Bowman, Alden, Wierwille, & Carroll, 2008a) to classify level of drowsiness.

An approach under development in the European Community is the "System for effective Assessment of driver vigilance and Warning According to traffic risK Estimation" (AWAKE) project (AWAKE, 2010). This program is aiming for an algorithm that provides at least 90 percent accuracy with less than a 1 percent false alarm rate. The algorithm proposed uses eye lid data, steering wheel grip and lane keeping, to classify the level of drowsiness as awake, may be drowsy, or drowsy. No detailed descriptions of the algorithm or results are currently available.

1.3.4. Recommendations

Existing drowsy driving detection algorithms can serve as benchmarks or points of comparison in the evaluation of the effectiveness of the IMPACT algorithms for the detection of drowsy driving. To warrant implementation and study, comparison algorithms must meet several criteria: They must be (1) sufficiently detailed and feasible to implement, (2) supported by evidence of their effectiveness, and (3) include different approaches using both individualized and generic algorithms.

Based on the criteria for this research, the most promising comparison algorithms for implementation are two related to PERCLOS (PERCLOS and PERCLOS+), and the steering behavior algorithm (King et al., 1998). Unlike many drowsiness detection algorithms, these algorithms meet the established criteria particularly related to sufficient detail for implementation. Additionally, they provide a driver-based, vehicle-based and combination algorithms focused on continuous detection of drowsiness that complement the event-based algorithms for detection of alcohol impairment that will be evaluated from the prior IMPACT work.

PERCLOS uses video of the driver's face to determine the proportion of time that the driver's eyes are more than 80 percent closed over a particular time window, sometimes as small as 1 minute. This algorithm is highly effective at identifying drowsy driving using a model of the individual's eyes to accurately detect proportion of eye closure. It is detailed sufficiently in the literature, generally accepted, and available commercially in many eye tracking systems including FaceLab.

PERCLOS+ combines vehicle-based measures and PERCLOS to identify drowsy drivers. The data needed to support this algorithm are easily accessible within the simulation environment. Early results show promise, although published data on the overall analysis of algorithm performance is not yet available (Hanowski, Bowman, Alden, Wierwille, &

Carroll, 2008b). This algorithm appears to use the combined data sources to improve the sensitivity and robustness of the PERCLOS algorithm.

King et al. (1998) proposed a purely vehicle-based algorithm using steering inputs that does not consider direct data about the drowsy driver state, such as eye closures. It has the potential to detect drowsiness relatively early because it considers degradation in steering control before it results in degraded lane keeping, such as lane departures used in the PERCLOS+ algorithm which risks misses if the driver is able to avoid departing the lane. Sufficient detail is available to implement the algorithm, as well as access to the data required to make the algorithm work. Another advantage is that it does not rely on PERCLOS, unlike the other two algorithms that will be compared to the IMPACT algorithms.

Other potential algorithms that were considered were not included for a variety of reasons. EEG- based algorithms have been found to be less reliable than the PERCLOS approach and would have required additional equipment and integration, the ePERCLOS algorithm appears similar in effectiveness to other algorithms, such as Mattsson (2007) and King et al. (1998), and is based on PERCLOS. The facial expression algorithm (Ji et al., 2004) did not provide sufficient details to implement and would likely have required additional hardware and software. Although they are not a promising algorithm input, because of the close association of EEG with sleep, EEG- based metrics are used in conjunction with other measures to identify drowsiness.

One of the aims of this effort is to consider the individualization of algorithms in predicting impairment. Individualization can be regarded in terms of measurement or in terms of thresholds. Individualization of measurement largely focuses on differences in how driver-based measures are captured, such as facial features or eye models. Individualization in thresholds for classification has been less used. Individualization of the threshold requires sufficient data in both the impaired and non-impaired state to properly train, which is difficult in a short experimental session, as well as on a road where driver state is difficult to accurately ascertain. For this reason, the focus was on selecting at least one algorithm that individualizes based upon driver features, while including other algorithms for which individualization of thresholds is feasible.

Three algorithms PERCLOS, PERCLOS+, and steering behavior were selected as the comparison algorithms. The PERCLOS and PERCLOS+ algorithms, both use individualization in their models of eye closure. The PERCLOS and steering behavior algorithms, both lend themselves to individualization of the thresholds, at which drivers are classified as drowsy.

2. DATA COLLECTION METHODS

Data were collected from drivers both while alert and while drowsy, during representative driving scenarios in a high-fidelity driving simulator. The following sections summarize the data collection methods: participant population, simulator and sensor suite, experimental design, procedure, and dependent variables.

2.1. Participants

Seventy-two participants[1] completed three drives: one during the daytime, one when moderately drowsy, and another when severely drowsy. The drivers were healthy men and women from three age groups (21to34, 38 to 51, and 55 to 68 years old). Each possessed a valid State-issued driver's license. Participants were paid $250 for completing all study sessions. Pro-rated compensation was provided for participants who did not complete the study.

Participants were recruited from the NADS Participant Database, Internet postings, and referrals (see Appendix B for recruitment material). An initial telephone interview determined eligibility for the study. Applicants were screened for health history, current health status (see Appendix C), and whether they were a morning or evening person (Adan & Almirall, 1991) (see Appendix D). To eliminate potential participants that were very awake during the overnight data collection periods, applicants with scores on the morning/evening scale less than 12 out of 30 were not eligible for participation. Those with scores indicating that they were an early morning person were not excluded. Pregnancy, disease, sleep disorders, or evidence of substance abuse resulted in exclusion from the study. Applicants taking prescription medications that cause or prevent drowsiness were also excluded from the study.

In particular, the criteria required that participants were licensed and drove at least 10,000 miles per year for the past 2 years, had no restrictions on their driver's license except for vision, were not currently taking illegal drugs or medications that cause or treat drowsiness, and had no warning signs for obstructive sleep apnea (Brown et al., 2009). They also had to live within a 30-minute drive to the National Advanced Driving Simulator (NADS), be able to participate after 7 p.m., stay awake overnight without sleeping, abstain from caffeine consumption after 12 p.m. on the day of overnight visit, and abstain from driving during the day following the overnight visit. In addition, participants needed to have sleep patterns that include going to bed and waking up at approximately the same time every day, not use any special equipment to drive, such as pedal extensions, hand brake or throttle, spinner wheel knobs, or other nonstandard equipment, and not have participated in distraction or alcohol and driving studies conducted at the NADS. Additional details on participant enrollment can be found in Appendix E.

2.2. Simulator and Sensor Suite

The NADS is located at the University of Iowa's Oakdale Campus. It consists of a 24-foot dome in which an entire car is mounted (see Figure 1). All participants drove the same vehicle—a 1996 Malibu sedan. The motion system on which the dome is mounted provides 400 square meters of horizontal and longitudinal travel, and ±330 degrees of rotation. Each of the three front projectors has a resolution of 1600 x 1200; the five rear projectors have a resolution of 1024 x 768. The edge blending between projectors is 5 degrees horizontal. The NADS produces a thorough record of vehicle state (e.g., lane position) and driver inputs (e.g., steering wheel position), sampled at 240 Hz.

Figure 1. Representation of NADS driving simulator (left) with a driving scene from inside the dome (right).

Figure 2. Face Lab cameras mounted in the Malibu cab with a separate head tracking system mounted between them.

The cab was equipped with a Face Lab 5.0 (Seeing Machines, Canberra, Australia) eye-tracking system that was mounted on the dash above the steering wheel. The worst-case head-pose accuracy was estimated to have RMS error of 5 degrees. In the best case, where the head was motionless and both eyes were visible, a fixated gaze may be measured with an estimated error of 2 degrees. The eye tracker records data at a rate of 60 Hz. The cab was also equipped with a Seeing Machines Driver State Sensor (DSS) V3.4.260101, a single-camera system that was used for head tracking. The installation of the cameras is shown in Figure 2.

The driver's seat was configured with a set of 14 seat sensors that provide posture data. This included six on the base of the seat with three running along each side, and eight on the back of the seat with four running along each side. Data from these sensors were collected at 60 Hz. They were not used for any of the drowsiness detection algorithms, but were needed for a distraction detection algorithm that will be examined in future research.

The study also used the B-alert X-10 to collect EEG data from F3,Fz, F4, C3, Cz, C4, P3, POz, and P4 and heart rate data (Advanced Brain Monitoring, 2011). These signals were used to generate proprietary metrics of task engagement, distraction, drowsiness, and workload to help validate the effectiveness of the experimental manipulations and will also be available for future research.

Additional sensors were used to ensure that participants followed the procedure. An Alco-Sensor IV (Intoximeters Inc., 2011) breath-alcohol-testing instrument was used to measure participants' breath alcohol concentration (BrAC). The hand-held sensor uses a fuel cell to determine BrAC. The system was checked at least every other day for calibration and recalibrated using an approved dry gas standard. A Motionlogger Actigraph (Ambulatory Monitoring Incorporated, 2009) was used to measure participants' activity level to determine when participants were sleeping for the two days prior to each visit.

2.3. Driving Scenarios

The scenarios were largely the same as those that were used in the IMPACT study (Lee et al., 2011). This scenario was selected as the starting point for the scenario for this study in order to provide continuity with prior driver impairment research examining alcohol and distraction. By keeping the driving environment and the driving events largely constant, it allows for future comparisons and algorithm development in Phase 2 of this research which will examine alcohol- impairment, distraction and drowsiness.

Each drive included three connected nighttime driving segments. The drives started with an urban segment composed of a two-lane roadway through a city with posted speed limits of 25 to 45 mph, as well as signal-controlled and uncontrolled intersections. An interstate segment followed that consisted of a four-lane divided expressway with a posted speed limit of 70 mph. After a period in which drivers followed the vehicle ahead, they made lane changes to pass several slower-moving trucks. While on the expressway, a CD changing task, consistent with that used in the IMPACT study.[2] The drives concluded with a rural segment featuring a two- lane undivided road with curves onto a gravel road. In a difference from the IMPACT study, the final segment of the drive included an extension of the original gravel roadway from IMPACT, and then a 300-second straight paved roadway. These three segments mimicked a drive home from an urban parking spot to a rural location via an interstate. Scenario events (driving segments with turns, signals, curves, interstate truck following, a dark rural road, etc.) in each of the three segments combined to provide a representative trip home of approximately 35 minutes, in which drivers encountered situations that might be encountered in a real drive. Throughout the urban section, a series of potential hazards required drivers to scan the roadside. These hazards included pedestrians, motorbikes, and cars entering and exiting the roadway. These hazards had paths that would cross the driver's path if they were to remain on their initial headings. There was an instance where a pedestrian crossed the driver's path well in front of the driver. Scenario events are summarized in Appendix F, Table 4. The differences from IMPACT are the extension of the drive to include additional time on the gravel roadway, a transition back to a paved road, and a ten minute drive on a straight roadway to end the drive instead of pulling into a driveway, as in the IMPACT scenario. These changes were implemented to improve sensitivity of the scenario to the effects of drowsiness, as discussed in Section 1.1, by adding a segment of drive that is most likely to be problematic for drowsy divers while maintaining the ability to compare back to prior data.

Each participant drove the simulator three times, once in a daytime alert condition, once in a moderately drowsy condition and once in a more severe drowsy condition. All three drives were completed with nighttime visual scene. Three scenarios with varied scenario event orders (but the same scenario events) were used to minimize learning effects from one drive to the next. Each of the three scenarios had the same number of curves and turns, but their order varied. For example, the position of the left turn in the urban section varied so that it was located at a different position for each drive. Additionally, the order of the left and right rural curves varied between drives. The scenario specification in Appendix F provides additional details concerning the differences among the three scenario event sequences.

2.4. Experimental Design and Independent Variables

A 2 x 2 x 3 x 3 mixed-design exposed 12 groups of participants to three drowsiness levels in two different orders. Between-subject independent variables were: age group, gender, and order of the drowsy and alert drives. The within-subject independent variables were drowsiness: (daytime) alert, (nighttime) moderate drowsiness and (nighttime) severe drowsiness, with two nighttime drowsiness sessions blocked into one visit, such that the moderate drowsy drive preceded the severe drowsy drive. The blocking of these two drives conforms to the natural pattern of increased drowsiness across an evening and is consistent with other prior studies looking at drowsiness in which repeated performance measures are collected across a single session. Although this blocking does have the potential to introduce a confound, this method was chosen to most closely replicate the natural process of increased drowsiness and because it avoids potential confounds associated with different amounts of continuous time awake if the overnight drives occurred separately.

2.4.1. Age and Gender

The choice of age range was made to match the data previously collected with alcohol impaired drivers in the IMPACT project. Three factors motivated the choice of the age ranges in that study. The first factor was that only those who could legally drink in Iowa would be included. Therefore, enrollment in the study was restricted to those 21 or older. The second factor was that to the extent practical, the entire spectrum of adults who drink and drive should be included, which motivated including the older age group. The third factor was that the age ranges should be uniform, with equal spacing between them. Thus, each group had a range of 14 years. Both male and female drivers were included in the study.

2.4.2. Drowsiness

The choice of the daytime alert and drowsy conditions was designed to provide data that are clearly differentiated. The daytime alert drive occurred during the morning (nominally) alert period between 9 a.m. and 12 noon. The nighttime drowsy drives began between 10 p.m. and 1 a.m. (moderately drowsy) and 2 a.m. and 5 a.m. (severely drowsy). The severely drowsy condition occurred after at least 18 hours of continuous wakefulness. The order of the daytime alert and nighttime drowsy conditions was counterbalanced to partially avoid confounding from learning effects.

Table 1. Participants assigned to each alertness sequence and scenario sequence

Alertness Sequence[1]	Driving Scenario Sequence[2]	Age					
		21-34		38-51		55-68	
		Gender		Gender		Gender	
		Male	Female	Male	Female	Male	Female
1	1	2	2	2	2	2	2
1	2	2	2	2	2	2	2
1	3	2	2	2	2	2	2
2	1	2	2	2	2	2	2
2	2	2	2	2	2	2	2
2	3	2	2	2	2	2	2
Total		12	12	12	12	12	12

Note. [1]Alertness sequence 1 = Alert, Drowsy; 2 = Drowsy, Alert.
[2]Driving Scenario Sequence 1 = Scenario A, B, C; 2 = Scenario B, C, A; 3 = C, A, B.

2.5. Procedure

Following a screening visit, each driver participated in three data collection sessions; two occurred during the night visit, which was separated by at least 3 days from the day visit. This differs from IMPACT, in which the three visits were 7 days apart. Order of visits (alertness sequence) and assignment to a scenario event sequence were counterbalanced across participants as shown in Table 1. A summary of the study procedures is found in Appendix G.

2.5.1. Screening Visit

On study Visit 1 (screening), each participant first gave informed consent to participate in the study and received a copy of the signed informed consent form (see Appendix H). They then provided urine samples for the drug screen and, for females, the pregnancy screens. The drug screen was a 10-panel test for amphetamines, methamphetamines, benzodiazepines, cocaine, marijuana, methadone, phencyclidine (PCP), barbiturates, tricyclic antidepressants, and morphine/opiates. Any other medications were reported by participants. Measurements of blood pressure and heart rate were then made. Cardiovascular measures within acceptable ranges (systolic blood pressure = 120 ± 30 mm Hg, diastolic blood pressure = 80 ± 20 mm Hg, heart rate = 70 ± 20 bpm) and a negative BrAC confirmed eligibility for the study. Eligible participants then completed a demographic survey that included questions related to crashes, moving violations, driver behavior, and driving history (see Appendix I). They viewed an orientation and training presentation (see Appendix J) that provided an overview of the simulator cab and the in-cab CD changing task they would be asked to complete while driving. Participants then completed an approximately 8-minute practice drive that included making a left-hand turn, driving on two- and four-lane roads, and practicing the CD changing task. They received recorded audio navigational instructions to guide them through the route. Appendix K describes the in-cab protocol that was administered. After the drive they completed a wellness survey that asks questions about how they felt (see Appendix L). If the survey indicated a propensity for simulator sickness based on total score greater than 35 or nausea scores greater than 40, the participant was ineligible to continue. If still eligible, the participant was fitted with a B-Alert cap and electrodes, and completed an EEG baseline procedure.

Two days prior to Visit 2, participants were given an activity monitor (Actigraph) that they were instructed to wear until Visit 3. It recorded periods of activity and sleep prior to their study visits. Participants also were instructed to keep a written activity log (see Appendix M) during this period to provide more details about activities that could affect their alertness.

2.5.2. Daytime-Alert Visit

Participants were asked to not ingest any caffeine on the days when they underwent their daytime alert condition. They drove themselves to the facility. Upon arrival, the activity monitor and activity log were collected and data uploaded and recorded. In addition to the activity log that the participants brought with them, they completed a survey that asked questions about their sleep and food intake over the past 24 hours, (see Appendix N). The monitor and log data were reviewed to ensure that the participants had a normal night's sleep (at least 6 hours) the preceding night. Their BACs were checked to ensure that they were not under the influence of alcohol (BAC of zero). Participants who did not meet the sleep or BAC requirements were dropped from the study. Participants were then fitted with the wireless B-Alert cap and electrodes to record their EEGs and heart rates. The participants then entered the simulator and eye tracking calibrations were completed.

Prior to beginning the drive, the participants also completed a questionnaire about their current sleepiness level, the Stanford Sleepiness Scale (Hoddes et al., 1973) (see Appendix O), and a version of the Psychomotor Vigilance Test or PVT (Cognitive Media Iowa City, IA) based on the Psychomotor Vigilance Task (Wilkinson & Houghton, 1982). This version of the PVT displayed a target to which the participant responded as quickly and accurately as possible by a button press. Although the duration of the PVT is generally 10 minutes, more recent research has supported the use of shorter duration tasks (Loh et al., 2004). This version of the test was implemented on an iPad, and provides both a 5 and 10-minute version for use at different times in the procedure. The participants drove through the simulation scenario after completing the 5- minute PVT in the vehicle.

Following the drive, participants were again administered the Stanford Sleepiness Scale), the wellness survey, PVT, plus a Retrospective Sleepiness Scale (See Appendix P) and a simulator realism survey (see Appendix Q). The Retrospective Sleepiness Scale required subjective judgments of drowsiness at specified scenario locations. The B-Alert cap was then removed. If the participants had not already completed their nighttime-drowsy visit, the activity monitor and activity log were returned to them and they were reminded of their next appointment.

2.5.3. Nighttime-Drowsy Visit

Participants were instructed to restrict beverage consumption to water after 12 p.m. on the day of their overnight visit, to minimize caffeine intake. They were provided with a list of items to avoid that contained caffeine including coffee, tea, soda, vitamin water, energy bars, energy drinks, and foods with chocolate. On nights when participants underwent their nighttime drowsy condition, they were picked up at their homes after having eaten dinner, and transported to the simulation facility to arrive around 7 p.m. Upon arrival, the activity monitor and activity log were collected and data recorded. While the data were being recorded, the participants completed sleep and food intake surveys. The activity monitor and log data were reviewed to ensure that the participants had a normal night's sleep (at least 6 hours) the preceding evening and did not take any naps during the day. If a participant indicated that the monitor was worn and the data were not recorded, only the log was used to determine if the participant was eligible to continue. If a participant indicated that the monitor was taken off or not worn, he or she was dropped for non-compliance to the protocol. Participants' BAC was checked to ensure that they were not under the influence of alcohol. Participants who did not meet the sleep or BAC requirements were dropped from the study and returned home. Each participant's caffeine intake was reviewed in the activity log and again in the sleep and intake log. If caffeine was consumed after noon on the day of the overnight drive, the participant was either rescheduled or dropped from the study. Participants were assigned to simulator drive times based on their waking times; therefore, based upon their survey responses and the activity logs, the participant who had awakened the earliest was selected to drive first and so on. Participants were then fitted with the B-Alert monitoring device.

A variety of activities were provided to keep participants awake including activities on an iPad, reading, playing computer games, etc. They were monitored to ensure they did not fall asleep or converse with other participants. If participants began to fall asleep, they were engaged by a researcher to keep them awake. The participants completed the Stanford Sleepiness Scale every 30 minutes until they drove. One hour prior to their drive, they were taken to a private room to wait. They completed a PVT at this time, and also at 30 minutes prior to the drive. Participants were escorted to the simulator between 10 p.m. and 1 a.m. for their first drives. Once in the simulator, eye tracking calibration procedures were performed, and the B-Alert electrode connection was verified. Before starting the drive, the participants completed a PVT and Stanford Sleepiness Scale. After the drive, participants completed the Stanford Sleepiness Scale, a Wellness Survey, a PVT, and a Retrospective Sleepiness Scale.

Participants were then escorted back to a separate waiting area where TV, movies, reading, computer games, etc. were available. A Stanford Sleepiness scale was administered every 30 minutes until their next drive. One hour prior to their second drive times, participants were again taken to a private room to wait. They completed a PVT one hour prior to the drive and also at 30 minutes prior to the drive. Participants were escorted to the simulator between 2 a.m. and 5:30 a.m. for their second drives. Once in the simulator, eye-tracking calibration procedures were performed, and the B-Alert connection was verified. Before starting the drive, the participants completed a PVT and Stanford Sleepiness Scale. After the drive, participants completed Stanford Sleepiness Scale, a Wellness Survey, a PVT, a retrospective sleepiness scale, and a realism survey. The B-Alert system was then removed. If the participants still needed to return for their daytime-alert visit, the activity monitor and activity log were returned to them, and they were reminded of their next appointment. At the end of their third visit, participants were given a debriefing survey, (see Appendix R) and paid $250. Then the participants were given the debriefing statement (see Appendix S) and driven home.

2.6. Dependent Variables

The dependent variables differed across the 22 distinct scenario events that comprised the three segments of the drive. The primary measures were lane position (mean, standard deviation, departures), speed (deviation from limit, standard deviation), steering (reversals, heading error), lateral acceleration (maximum, jerk rate), eye closure (blinks), head position (standard deviation). The scenario specification describes the dependent variables for each scenario event (see Appendix F). Potential intervening variables and their mitigation are discussed in Appendix T.

2.6.1. Data Verification and Validation

The data reduction began with verification of the raw input data. The data was then aggregated as needed to support sensitivity analyses and algorithm development and testing. The process concluded with validation of metrics that summarize the data.

Verification concerns the process of ensuring that the raw data accurately reflected the state of the vehicle, driver, and roadway. Scenario event errors, database flaws, and measurement noise all may contribute to spurious raw data that would need to be removed before they are transformed into measures of driver behavior. Several automatic data checks combined with manual visualizations identified these issues. The verification procedures included verifying that all the variables in the raw data contain values, and that the file was of the expected size. The integrity of each variable was assessed on three factors: whether the values lie within the expected range, whether the values vary in a meaningful manner, and whether the variation in the values was continuous. These three indicators were automatically assessed or revealed in a plot of the data.

Validation concerns the process of ensuring that the summary measures accurately reflect the driver behavior or vehicle performance of interest. Measures based on aggregating measures at the sample level (across scenario events, drives, or people) might fail to reflect the underlying population differences in behavior due to such issues as differences in the distribution of the data, or the presence of data that differs in significant ways from the rest of the sample. Data visualization techniques provide a useful tool for addressing challenges to the validity of such summary measures by examining them in the context of the time history and distribution of the data. Data were visualized by superimposing the summary measures over the raw data with a reference point, such as the posted speed limit, to roughly assess whether the underlying calculations are correct and in fact capture the behavior of interest, as opposed to separate types of behavior that might otherwise have been combined.

In the following section, data will be reviewed and analyzed to assess the sensitivity of the measures to the drowsiness manipulation. This section will include an analysis of PERCLOS, metrics derived from EEG and heart rate, driving performance measures, PVT, and self-reports of sleepiness. The analysis will focus on documenting patterns of performance that differentiate the three levels of drowsiness over the drive. The analysis will also consider the how well measures taken outside the drive, (PVT, self-reported sleepiness, and hours of wakefulness) predict measures obtained during the drive.

3. LEVEL OF DROWSINESS AND DRIVING PERFORMANCE

3.1. Drowsiness of Participants

Table 2 reports the cumulative time awake (CTA, in minutes) for the three drowsiness conditions: day, early night, and late night. As expected, the greatest CTA was measured in the late night condition (1,230 min), followed by the early night condition (1,001 min), and the day condition (222 min).

Table 3 reports the SSS scores that were obtained pre-drive, post-drive, and the averages for the three drowsiness conditions. The SSS has a range of 1 to 7 with 1 feeling active, vital, alert or wide awake and 7 being no longer fighting sleep, sleep onset soon having dreamlike thoughts. The average sleepiness score for the day drive was 2.35. The average sleepiness score for the early night and late night drives were 3.77 and 5.19, respectively. Thus, the highest level of sleepiness was measured for the late night drive, followed by the early night drive and the day drive. Note that in some cases the scale was not administered, resulting is some missing data.

Table 4 reports the pre-drive, post-drive, and average for the psychomotor vigilance test (PVT) across the three drowsiness conditions. The average PVT reaction time for the day drive was 382 ms. The average PVT reaction time for the early night and late night drives was 404 ms and 445 ms, respectively.

Table 2. Average cumulative time awake by drowsiness condition

	Day				Early Night				Late Night			
	N	*M*	*SD*	Median	*N*	*M*	*SD*	Median	*N*	*M*	*SD*	Median
CTA	72	223	73	214	72	1,001	53	995	72	1,230	51	1,228

Table 3. Average Stanford Sleepiness Scale scores by drowsiness condition

	Day				Early Night				Late Night			
Measurement	*N*	*M*	*SD*	Median	*N*	*M*	*SD*	Median	*N*	*M*	*SD*	Median
Pre-drive	68	1.8	.8	2.0	69	3.4	1.2	3.0	72	5.0	1.3	5.0
Post-drive	71	2.9	1.2	3.0	68	4.1	1.3	4.0	71	5.4	1.3	6.0
Average	68	2.4	.9	2.0	65	3.8	1.2	4.0	71	5.2	1.2	5.5

Table 4. Average psychomotor vigilance reaction times (ms) by drowsiness condition

	Day				Early Night				Late Night			
Measurement	*N*	*M*	*SD*	Median	*N*	*M*	*SD*	Median	*N*	*M*	*SD*	Median
Pre-drive	72	371	44	364	72	397	53	386	72	430	62	419
Post-drive	72	394	52	387	72	412	58	414	72	460	74	448
Average	72	382	46	379	72	404	52	400	72	445	62	441

Table 5. Pearson correlations between testing drowsiness condition (time of day) and selected measures of sleepiness

Measure	1	2	3	4
1. Drowsiness Condition				
2. Cumulative Time Awake (CTA)	.95			
3. Stanford Sleepiness Scale (Pre/Post Average)	.73	.69		
4. Psychomotor Vigilance Test (pre/Post Average)	.43	.39	.47	

Table 5 reports the correlations between CTA, SSS average, and PVT reaction time average. Pearson correlations ranged from .394 to .949. (all significant at the .01 level). The pattern of correlation sizes indicates that CTA-SSS and CTA-PVT correlations varied in size. This suggests that measures of sleepiness did not vary solely as a function of time awake since last sleep, but potentially also as a function of time of day, circadian rhythms, and possibly the participants' level of arousal during the entire test session at the NADS.

3.2. Driver Adaptation to Scenario Events with Repeated Exposure

The effect of repeated exposure was examined for lane deviation, mean speed, and speed deviation to determine if there was a systematic change across sessions. Analyses of variance with alpha level set at .05 were used to determine whether there were reliable differences as a function of session. No efforts were made to control for the family-wise Type I error. There were 12 scenario events for which lane deviation showed a significant difference across sessions. Only the gravel rural extension showed a pattern of improved performance across visits. There were 10 scenario events for which average speed showed a significant difference across scenario events. For all but one of those scenario events, there was an increase in average speed; however, the increase in average speed was less than 4 mph, at its greatest. There were 6 scenario events for which speed deviation showed a significant difference across visits. For all but one of those scenario events, there was a decrease in variability across visits. Twenty-one of the 28 significant differences were associated with short scenario events lasting approximately 30 seconds or less. For 60 of the 75 comparisons, there was not a significant pattern of learning observed. Overall most scenario events across these variables did not demonstrate a systematic learning effect or adaptation by the driver across visits. More detail can be found in Appendix U.

3.3. Effect of Drowsiness on Driving Performance Across Roadway Conditions

3.3.1. Analysis of Variance
A 2 x 3 x 3 between-between-within ANOVA was performed on each of the three composite measures for lane deviation, average speed and speed deviation. The composite scores were calculated by averaging the z scores of each measure across the scenario events

and by re-standardizing the mean into a T-score ($M = 50$, $SD = 10$). Additional details on the individual scenario events and the composite measure can be found in Appendix V. Between-subjects independent measures were gender and age group (21 to 34, 38 to 51, 55 to 68). Within-subjects independent measure was drowsiness condition (day, early night, and late night).

Table 6. Lane deviation composite score by drowsiness condition, age group, and gender

Age Group	Gender		Day	Early Night	Late Night
21-34	Females	M	50.43	48.62	53.68
		N	12	12	12
		SD	9.64	7.82	9.96
	Males	M	46.84	44.47	51.66
		N	12	12	12
		SD	9.86	7.68	10.98
	Total	M	48.63	46.54	52.67
		N	24	24	24
		SD	9.71	7.87	10.30
38-51	Females	M	52.21	49.24	52.96
		N	12	12	12
		SD	13.11	14.29	16.81
	Males	M	50.73	48.87	52.57
		N	12	12	12
		SD	10.16	8.11	13.60
	Total	M	51.47	49.06	52.77
		N	24	24	24
		SD	11.50	11.36	14.96
55-68	Females	M	49.52	48.14	53.32
		N	12	12	12
		SD	5.29	7.40	7.51
	Males	M	48.70	47.43	50.63
		N	12	12	12
		SD	7.47	6.82	7.46
	Total	M	49.11	47.78	51.97
		N	24	24	24
		SD	6.34	6.97	7.45
Total	Females	M	50.72	48.67	53.32
		N	36	36	36
		SD	9.66	10.04	11.74
	Males	M	48.76	46.92	51.62
		N	36	36	36
		SD	9.12	7.57	10.69
	Total	M	49.74	47.79	52.47
		N	72	72	72
		SD	9.38	8.87	11.18

3.3.1.1. Lane Deviation Composite Scores

The mean lane deviation composite scores by drowsiness condition, age group, and gender are shown in Table 6. Mauchly's test of sphericity was not significant, indicating that no adjustment to the degrees of freedom was required. Drowsiness condition produced a statistically significant main effect $F (2, 132) = 15.22$, $p < .001$, partial $\eta^2 = .19$. As shown in Figure 3, lane deviation composite scores varied as a function of drowsiness condition $F (1, 66) = 9.28$, $p < .01$, partial $\eta^2 = .12$. As shown in Table 7, lane deviation was not statistically different between the day and the early night conditions, and between the day and late night conditions. It was, however, statistically different between the early night and the late night conditions.

3.3.1.2. Average Speed Composite Scores

The mean speed composite scores by drowsiness condition, age group, and gender are shown in. Mauchly's test of sphericity was statistically significant, and the Greenhouse-Geisser adjustment was used to adjust the degrees of freedom. Drowsiness condition produced a statistically significant main effect, $F (1.64, 107.99) = 13.19$, $p < .001$, partial $\eta^2 = .17$. As shown in Figure 4, average speed composite scores varied as a function of drowsiness condition.

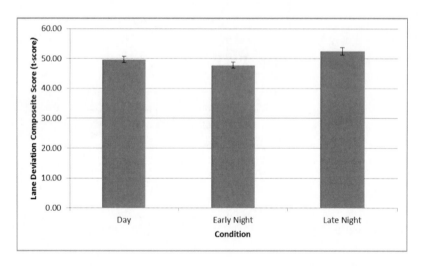

Figure 3. Lane deviation composite score as a function of drowsiness condition. Error bars represent standard error.

Table 7. Post-hoc Lane Deviation Comparisons

Comparison		Mean Difference	Std. Error	Significance	99.9% Confidence Interval for Difference	
					Lower Bound	Upper Bound
Day	Early night	1.95	.77	No	-.97	4.87
Day	Late Night	-2.73	.90	No	-6.13	.66
Early night	Late night	-4.67	.88	Yes	-8.00	-1.35

Note. Pairwise comparisons were conducted with α=.001.

As shown in Table 9, there was a statistically significant difference in average speed between the day and the early night conditions, but not between the day and late night conditions and between the early night and late night conditions.

Table 8. Average speed composite score by drowsiness condition, age group, and gender

Age Group	Gender		Day	Early Night	Late Night
21-34	Females	M	55.45	50.43	53.78
		N	12	12	12
		SD	6.73	9.44	7.94
	Males	M	59.15	57.82	59.32
		N	12	12	12
		SD	8.54	6.57	9.03
	Total	M	57.30	54.13	56.55
		N	24	24	24
		SD	7.75	8.80	8.78
38-51	Females	M	49.54	45.34	47.18
		N	12	12	12
		SD	9.17	11.26	11.57
	Males	M	56.71	51.29	52.84
		N	12	12	12
		SD	6.75	5.86	6.33
	Total	M	53.12	48.31	50.01
		N	24	24	24
		SD	8.69	9.29	9.57
55-68	Females	M	46.31	41.06	42.07
		N	12	12	12
		SD	7.14	8.00	8.53
	Males	M	43.71	43.02	45.00
		N	12	12	12
		SD	8.61	9.10	7.88
	Total	M	45.01	42.04	43.53
		N	24	24	24
		SD	7.85	8.44	8.17
Total	Females	M	50.43	45.61	47.67
		N	36	36	36
		SD	8.45	10.15	10.41
	Males	M	53.19	50.71	52.38
		N	36	36	36
		SD	10.38	9.39	9.64
	Total	M	51.81	48.16	50.03
		N	72	72	72
		SD	9.50	10.04	10.24

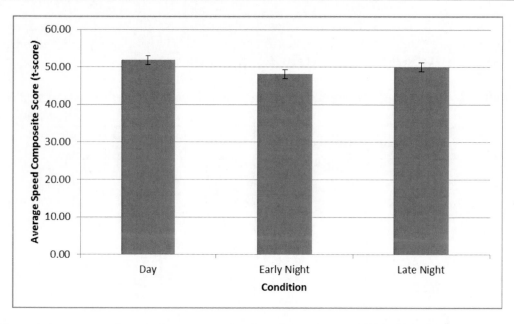

Figure 4. Average speed composite score as a function of drowsiness condition. Error bars represent standard error.

Table 9. Post-hoc comparison for average speed

Comparison		Mean Difference	Std. Error	Significance	99.9% Confidence Interval for Difference	
					Lower Bound	Upper Bound
Day	Early night	3.65	.80	Yes	.61	6.69
Day	Late Night	1.78	.78	No	-1.16	4.73
Early night	Late night	-1.87	.52	No	-3.83	.09

Note. Pairwise comparisons were conducted with α=.001.

3.3.1.3. Speed Deviation Composite Score

The mean speed deviation composite scores by drowsiness condition, age group, and gender are shown in Table 10. Mauchly's test of sphericity was not significant, indicating that no adjustment to the degrees of freedom was required. Drowsiness condition was not statistically significant.

3.4. Robustness of Metrics with Respect to Age, and Gender

3.4.1. Lane Departure Composite Scores

Although there was a significant effect of drowsiness condition on lane deviation, there were no effects for lane deviation relative to age and gender. There were no interactive effects between age and gender with drowsiness condition.

**Table 10. Speed deviation composite score by drowsiness condition,
age group, and gender**

Age Group	Gender		Day	Early Night	Late Night
21-34	Females	M	51.34	53.09	51.70
		N	12	12	12
		SD	11.43	15.00	10.06
	Males	M	44.11	45.19	48.20
		N	12	12	12
		SD	8.42	8.48	7.77
	Total	M	47.72	49.14	49.95
		N	24	24	24
		SD	10.49	12.58	8.97
38-51	Females	M	50.54	50.34	48.55
		N	12	12	12
		SD	5.97	7.49	6.82
	Males	M	49.17	50.86	49.71
		N	12	12	12
		SD	7.60	11.34	8.58
	Total	M	49.85	50.60	49.13
		N	24	24	24
		SD	6.72	9.40	7.60
55-68	Females	M	55.13	51.02	56.05
		N	12	12	12
		SD	14.36	12.61	9.29
	Males	M	49.88	47.44	47.70
		N	12	12	12
		SD	10.56	6.57	10.99
	Total	M	52.51	49.23	51.88
		N	24	24	24
		SD	12.61	10.00	10.83
Total	Females	M	52.34	51.48	52.10
		N	36	36	36
		SD	11.01	11.82	9.12
	Males	M	47.72	47.83	48.53
		N	36	36	36
		SD	9.07	9.07	8.99
	Total	M	50.03	49.66	50.32
		N	72	72	72
		SD	10.28	10.62	9.17

3.4.2. Average Speed Composite Scores

There was one effect on average speed relative to age. There was a significant main effect of age, $F(1, 66) = 16.08$, $p < .001$, partial $\eta^2 = .33$.

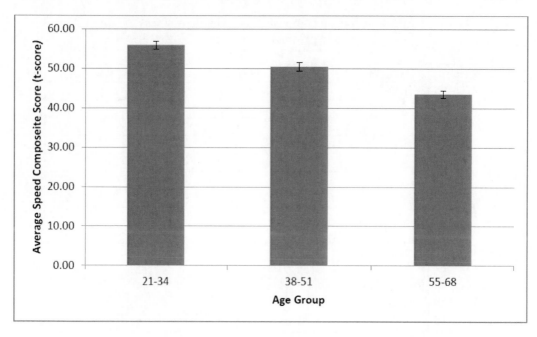

Figure 5. Average speed composite score as a function of age group. Error bars represent standard error.

Table 11. Post-hoc comparison of average speed for age

	Comparison	Mean Difference	Std. Error	Significance	99.9% Confidence Interval for Difference	
					Lower Bound	Upper Bound
21-34	38-51	5.51	2.20	No	-2.83	13.85
21-34	55-68	12.46	2.20	Yes	4.12	20.80
38-51	55-68	6.95	2.20	No	-1.39	15.30

Note. Pairwise comparisons were conducted with α=.001.

Figure 5 shows that there was a statistically significant difference in average speed between the 21-to-34 and the 55-to-68 age groups, but not between the 21-to-34 and the 38-to-51 groups and the 38-to-51 and 55-to-68 age groups. There were no interactive effects between age and gender with drowsiness condition.

3.4.3. Speed Deviation Composite Scores

There were no effects for speed deviation relative to age and gender. There were no interactive effects between age and gender with drowsiness condition.

3.5. Discussion

Drowsiness, as defined by the experimental conditions showed an effect for both lane deviation and average speed. The overall pattern of lane keeping was worse for the late night condition relative to the early night. The general pattern of the average speed was a decrease from the daytime drive to the early night drive and an increase with the late night drive, with only the difference in average speed between the early night and daytime speeds reaching statistical significance. For neither measure was there a systematic decrease in performance associated with an increase in drowsiness. The U- shaped pattern of performance indicates a more complex response to drowsiness where performance, particularly related to lane keeping, improves to a point before degrading, suggesting compensatory behavior as drivers respond to increased drowsiness. These results suggest that drowsiness does not follow a simple dose response relationship, with performance decreasing with increasing periods without sleep. However, the results also show lane keeping performance degrades the most in the situation where degraded performance is expected: late at night after a long period without sleep. The study succeeded in inducing drowsy driving.

4. ALGORITHM EVALUATION PLAN

The development of algorithms to detect drowsy driving is a topic of great interest to NHTSA and researchers around the world. Because drowsiness undermines driving safety, such algorithms could help reduce crashes and fatalities on U.S. highways. This section describes a process to assess algorithm effectiveness; but also considers the larger question of whether the algorithms can differentiate between drowsiness and other types of impairment detection, such as distraction and alcohol intoxication.

The algorithm development and evaluation relies heavily on previous research conducted for NHTSA concerning alcohol (Lee et al., 2010) and distraction (Lee et al., in review) impairment detection. Indeed, this study used a similar experimental protocol, scenarios, and data reduction process to maximize the opportunity for cross-study data comparisons.

This section describes an evaluation plan for drowsy driving detection algorithms. First algorithms that have been adapted from previous work or conceived as part of this study are described. Next, the criteria that were used to analyze algorithm effectiveness are presented, and then the steps of the evaluation are explained.

4.3. Impairment Detection Algorithms

This analysis considers several algorithms that have been selected for detecting drowsy driving. Some have been adopted from previous NHTSA studies, where the goal was to detect alcohol impairment and distraction (Lee et al., 2010; Lee et al., in review), while others have been added for their demonstrated sensitivity to drowsiness, such as PERCLOS and PERCLOS+ (Dinges, Mallis, Maislin, & Powell, 1998; Wierwille et al., 1994; Tijerina et al., 1999; Hanowski, Bowman, Alden, Wierwille, & Carroll, 2008). Thus, algorithms designed to detect various types of impairment were used to detect drowsiness. Assessing how algorithms tailored to detect specific impairments (i.e., alcohol, distraction, and drowsiness) perform in detecting drowsiness is one step toward assessing the degree to which a single algorithm might detect a range of impairments.

Algorithms able to detect a range of impairments are denoted as general, and those that detect single impairments are denoted as specific. A *specific* algorithm is one that has been developed to detect one type of impairment and might not be sensitive to other impairments. A *general* algorithm is designed to detect multiple types of impairment. A general algorithm may have been developed for one particular type of impairment and later expanded to fulfill a larger role. The ability of a general algorithm to succeed depends in part on the physiological and psychological similarity of the impairment mechanisms.

Alcohol acts as a central nervous system (CNS) depressant (Arnedt et al., 2001), and so one might expect drowsiness to exhibit similar influences on driving performance. In contrast, cognitive distraction loads working memory, and interferes with attention allocation, as manifested in gaze concentration (Regan, Lee, & Young, 2009). Drowsiness impacts cognitive ability and working memory as measured in psychomotor vigilance test, and results in microsleeps and more frequent eye closures. It is possible to counteract drowsiness to a certain extent with increased compensatory effort, but only up to a point (Kloss, Szuba, & Dinges, 2002). Drowsiness may share features with distraction in that the onset of symptoms may be relatively sudden and transient. Drowsiness may induce gaze concentration similar to distraction. Drowsiness would be expected to share some features with alcohol impairment as they both impact the CNS; indeed, the performance under the two types of impairments have been equated in several studies (Williamson & Feyer, 2000; Dawson, Drew, & Reid, 1997; Arnedt et al., 2001).

The NHTSA distraction detection and mitigation project (Lee et al., in review) considered visual and cognitive distraction. Four algorithms were implemented and evaluated for this project. Only one of the four was designed to detect cognitive distraction, which is included in this study because cognitive distraction may share characteristics of drowsy driving, namely a lack of active visual scanning of the forward scene signified by gaze concentration.

A truly general algorithm could help protect drivers from impairments not anticipated by the designer. This is a motivating factor in adapting proven alcohol and distraction algorithms for application to drowsiness in this study.

We considered algorithms applied at three distinct timescales, summarized in Table 12. The utility of an algorithm varies according to its timescale, with long range approaches being appropriate for post-drive evaluation, medium range ones appropriate for moderately spaced countermeasures, and short range for the detection of safety-critical situations.

The previous NHTSA study of alcohol (Lee et al., 2010) produced three algorithms that were sensitive to differences between the baseline condition and two BACs (.05 and .10 g/dL). These algorithms were based on logistic regression, boosted decision trees, and support vector machines (SVMs). Various measures of driver performance, environmental demand, and event type were used as inputs to the algorithms; and they were trained and tested on simulator data. The BAC classifications were grouped by scenario event because driver behavior during a yellow light dilemma, for instance, could vary considerably from that observed during highway driving. A decision tree algorithm with boosting was able to detect impairment with greater accuracy than the other candidates: support vector machines and logistic regression. For this reason, the event- based decision tree algorithm is one of the candidates evaluated in the current work to detect drowsiness.

Table 12. Three levels of algorithm timescale

Timescale	Description	Period	Indicators
Long range	Whole drive	~30 minutes	Stanford Sleepiness Score; Condition
Medium range	Event-based	~1-6 minutes	Retrospective Sleepiness Score
Short range	Real-time	~60 seconds	Drowsy lane departures

Table 13. Impairment Detection Algorithm Summary

Label	Algorithm	Source	Impairment	Timescale
PC	PERCLOS	(Dinges, Mallis, Maislin, & Powell, 1998)	Drowsiness	Medium
PC+	PERCLOS+	(Hanowski, Bowman, Alden, Wierwille, & Carroll , 2008)	Drowsiness	Medium
SB	Steering-Based	(King, Mumford, & Siegmund, 1998)	Drowsiness	Short
EEG	EEG	NHTSA DRIIVE	Drowsiness	Short
DT	Decision Tree	NHTSA IMPACT	Alcohol, Generalized	Medium
MDD	Multi-Distraction Detection	NHTSA distraction detection and mitigation	Distraction	Short
TLC	Time-to-lane-crossing	NHTSA DRIIVE	Drowsiness	Short
SRF	Steering random forest	NHTSA DRIIVE	Drowsiness	Short
BN	Bayes net	NHTSA DRIIVE	Alcohol, Generalized	Short

Additionally, the current project has developed a real-time algorithm to detect drowsiness that was trained and testing using data from the IMPACT study and applied to the drowsiness data. The new algorithm uses a Bayesian network to model the conditional probabilities associated with several driving performance measures.

A summary of the various algorithms is given in Table 13. Each of these algorithms was developed to detect a specific impairment, with several being developed specifically to detect drowsiness. This study assesses whether any of the alcohol-specific algorithms can also detect drowsiness as well as those developed specifically to detect drowsiness, and therefore offer promise as general algorithm that can detect and distinguish a wide range of impairments.

4.4. Algorithm Performance Criteria

Assessing algori1thm performance depends on comparing the classification (i.e., drowsy or alert) to the actual state of the driver. The actual state of the driver is sometimes referred to as the ground truth, and is ideally indicated by a "gold standard" measure that provides an unambiguous indicator of the driver state. Such a gold standard is difficult to define for drowsiness. Arguably a clinical EEG record scored by a sleep expert is the gold standard indicator of drowsiness, but it was not possible to obtain this indicator for this study. Instead, this study used several drowsiness measures, which combined to provide ground truth indicators. The measures relied on for this purpose included the pre-drive and post-drive SSS, pre-drive and post-drive PVT, and retrospective SSS (RSS). To assess algorithm timeliness, drowsiness-related lane departures represented the ground truth indicator of drowsiness and location-matched periods of alert driving represented the ground truth indicator of alertness.

Three standard criteria were used to assess algorithm performance in detecting and distinguishing impairments: accuracy, positive predictive performance (PPP), and area under curve (AUC). Accuracy measures the percent of cases that were correctly classified, while PPP measures the degree to which those drivers that were judged to be drowsy were actually drowsy. An algorithm can correctly identify all instances of impairment simply by setting a very low decision criterion, but such an algorithm would misclassify all cases where there was no impairment. The relationship between the true positive detection rate (sensitivity) and false positive detection rate (1-specificity) is represented by the receiver operator characteristic (ROC) curve. ROC curves are presented for many of the algorithm results. AUC represents the area under the receiver operator curve, which provides a robust and simple performance measure. Perfect classification performance is indicated by an AUC of 1.0, and chance performance is indicated by .50. AUC is an unbiased measure of algorithm performance, but accuracy and PPP are more easily interpreted, so all three are used in describing the algorithms.

Beyond the standard measures of algorithm performance, this study also considered the degree to which the algorithm offers a timely detection of impairment. Timeliness is most relevant to concurrent algorithms, which run in real-time and support time-critical warnings. In contrast, post-drive algorithms aggregate data over the length of the drive to provide post-drive feedback. An intermediate approach is exemplified by the IMPACT algorithms and could be called post-*event*, or event-based. For real-time algorithms, timeliness represents a critical performance metric that is likely to be balanced by accuracy—accumulating more data generally increases accuracy but undermines timeliness. To some extent, timeliness depends on the type of algorithm—some algorithms do not provide real-time indication of impairment due to the amount of data aggregation they require.

For those algorithms designed to produce real-time alerts, timeliness, the degree the algorithm can correctly detect impairment in advance of an impairment-related mishap, is added. For this analysis, timeliness is defined as its AUC six seconds before a drowsiness-related mishap, such as a drowsy lane departure. The locations of unintentional lane departures were determined during data reduction, and drowsy lane departures were verified by video review. It was expected that real-time algorithms would provide more accurate and timely drowsiness detection compared to algorithms that aggregate data across scenario events.

4.5. Research Questions and Hypotheses

Can algorithms designed for alcohol impairment detection (event-based decision tree, Bayes net) and distraction also detect drowsiness? Commonalities in the physiological basis of the impairments may cause drivers' performance to degrade in similar ways. Can algorithms designed for alcohol impairment detection be generalized to work well for both alcohol and drowsiness? Alcohol algorithms that have been retrained on drowsy driver data or new algorithms that include variables appropriate for drowsiness should be more accurate in detecting drowsiness than specialized alcohol detection algorithms, but may also be less accurate in detecting alcohol impairment if alcohol and drowsiness do share symptoms.

Can algorithms distinguish between alcohol and drowsiness-related impairment?

Do real-time algorithms perform better in detecting drowsiness in advance of a drowsiness-related mishap? Event-based algorithms, such as the decision tree algorithms previously used to detect alcohol impairment, may be less likely to have a high AUC value six seconds before the onset of a drowsiness-related mishap compared to a real- time algorithm.

4.6. Evaluation Method

4.6.1. Algorithms

Impairment detection algorithms can be characterized by the timescale over which they operate, and the timescale over which impairment indicators are expected to vary. Table 12 and Table 13 above present three distinct timescales that the algorithms use. The timescale assignments in Table 13 are not fixed. One may accumulate short or medium range algorithm outputs over a longer timeframe for a post-drive review for instance. Alternatively, one may attempt to sample medium range algorithms more often for real- time prediction, though the accuracy may suffer.

Other dimensions that separate the algorithms are the types of inputs they use (physiological or driving performance) and how the inputs are combined. Beyond these dimensions, some algorithms may be parametrically modified to become more general, perhaps by simply changing a parameter threshold. Alternatively, the more complicated alcohol algorithms may be retrained to a dataset obtained from drowsy driving, or a combined dataset consisting of both drowsy and alcohol impairments. Table 14 again lists the impairment detection algorithms that were used in this study, this time with inputs and outputs described.

Most of the algorithms produce a binary classification, making it the common basis for comparison between all the algorithms. In cases where an algorithm outputs something other than a binary output, the categorical or continuous outputs were mapped to a binary classification. Binary classifiers were obtained from more complex ones by setting thresholds. The details of obtaining a binary classification for drowsiness are given in the next section.

For each algorithm in Table 14, a binary output was created if one did not exist. Then the accuracy, PPP, AUC, and timeliness of each algorithm were calculated. These data were organized into two datasets: one based on scenario events and the other based on fixed windows of time with some percentage of overlap.

4.6.2. Driver data and drowsiness identification

Two datasets were created: event-based and continuous. The event-based data set follows the same format used in the IMPACT study, with the driving summarized in terms of 22- to 24-scenario events that range from about 6 to 680 seconds. The continuous data consists of driver and vehicle data recorded at 60 Hz for the entire drive. The continuous dataset was analyzed by organizing the data into time windows of a fixed time with some percentage of overlap. Each record of these datasets were coded as alert or drowsy according to three definitions: the drowsiness condition, a linear combination of PVT, pre-post and retrospective SSS, and the presence or absence of a drowsiness- related mishap. To maintain balance in the model training process, each data set was divided into equal numbers of drowsiness and alert instances.

4.6.3. Algorithm performance summary

Ten-fold cross validation was used to assess each algorithm, producing a measure of accuracy, PPP, AUC, timeliness and corresponding confidence interval for each algorithm. ROC curves were also used to summarize sensitivity and specificity graphically. In combination, these metrics were used to identify better or worse algorithms, and also to identify how they might complement each other. For example, some algorithms might not be timely, but they might be accurate. The optimal tradeoff between these factors remains an open question.

4.6.4. Algorithm generalization

Based on the results of this analysis, candidate algorithms for other target impairments were selected for adaptation to drowsy driving. The parameters were optimized for the drowsy data set, either through AUC analysis or re-training. Such changes to the parameters would undermine the ability of the algorithms to detect the impairment that they were originally designed to detect. The modified algorithms were analyzed and compared to the original ones. Potential generalizations of algorithms are considered as well. One method of generalization is simply to combine multiple specialized algorithms into one package.

Table 14. Impairment Detection Algorithm Inputs and Outputs.

Label	Algorithm	Inputs	Outputs
PC	PERCLOS	Eye closure	Continuous percentage Drowsy binary
PC+	PERCLOS+	Eye closure, lane departure	Drowsy categorical (low, moderate, severe)
SB	Steering-Based	Steering angle, steering rate	Drowsy binary
EEG	EEG	Scalp electrical activity	Continuous probability Drowsy binary
DT	Decision Tree	Multiple measures of driver performance	Intoxicated binary
MDD	Multi-Distraction Detection	Eye gaze location	Continuous PRC Visual binary Cognitive binary
TLC	Time-to-lane-crossing	Lane position, lane heading angle	Drowsy binary
SRF	Steering random forest	Steering wheel angle	Drowsy binary
BN	Bayes net	Multiple measures of driver performance, eye closure, eye closure rate	Intoxicated categorical (none, moderate, severe)

5. ANALYSIS OF ALGORITHM PERFORMANCE

This section describes algorithms and their ability to detect driver drowsiness. Similar algorithms have been developed to detect alcohol impairment (Lee et al., 2010) and distraction, and the central aim of this study is to assess how well these techniques can be used to detect drowsiness. The degree to which similar algorithms can detect both alcohol impairment and drowsiness, and the degree to which such algorithms can differentiate the two impairments, depends on the profile of the impairment over time and the particular manner in which the impairment influences driver behavior. Specifically, the impairment of alcohol is relatively constant over a period of 20 to 30 minutes and strongly influences lane keeping performance, whereas drowsiness might vary considerably over this period and might influence other elements of driving performance. These underlying differences in the profiles of impairment demonstrate the demands of developing algorithms to detect impairment. This study addresses the understanding of the demands of drowsiness detection by addressing the following questions:

- Can algorithms designed to detect alcohol impairment or distraction also detect drowsiness?
- Can algorithms designed to detect alcohol impairment be generalized to detect both alcohol and drowsiness?
- Can algorithms distinguish between alcohol and drowsiness-related impairment?
- Do real-time algorithms perform better than event-based or post-drive algorithms in detecting drowsiness in advance of a drowsiness-related mishap?

In order to answer these questions, several types of drowsiness measurement are used throughout the section. Each has its own merit and appropriate usage. SSS is a scale from one to 8 where one is alert and 8 is asleep. It was collected both pre and post-drive through a survey. The retrospective sleepiness scale (RSS) uses the same scale as SSS, and is administered via survey, but is an estimate from a continuous time measurement over the course of the drive. The psychomotor vigilance test (PVT) is an active memory test known to correlate with drowsiness. A 5-minute PVT was administered before and after each drive. A video review of lane departures was conducted to obtain a good quality set of truly drowsy scenario events against which to judge algorithm performance. The three timescales considered are summarized in Table 15, reproduced from Chapter 4.

SSS ratings and PVT scores are only appropriate when considering data from entire drives, while RSS data can be used for finer grain analysis or grouped at the scenario event or drive level. Drowsy lane departures are reliable events to compare against, but are transient in nature and not associated with drives or scenario events. Note that it is difficult to standardize terminology around the word drowsiness because the standard survey instruments used in this study use the word sleepiness. Throughout this section the terms drowsy and sleepy are used interchangeably.

Table 15. Three algorithm timescales

Aggregation	Description	Period	Indicators
Long range	Whole drive	~20-30 minutes	Post-drive SSS; Condition
Medium range	Event-based	~1-6 minutes	Event-based RSS
Short range	Real-time	~60 seconds	Drowsy lane departures

The following analyses address the research questions by first describing the distribution of drowsiness across drivers, conditions, and the drive. Drowsiness is classified here using a threshold of post-SSS rating greater than three. This distribution of drowsiness suggests algorithms used to detect alcohol impairment over the course of a 20-minute drive might perform relatively poorly, which is confirmed with an analysis of algorithms detecting impairment over the drive. The differences in the profiles of alcohol impairment and drowsiness are then used to create algorithms that detect alcohol impairment, drowsiness impairment and differentiate between the two. Real-time algorithms that aim to predict drowsiness associated with lane departures in advance of the lane departure are then considered. For that analysis, a more complex classification of drowsiness that combined SSS, RSS, PVT, and drowsy lane departures was used.

5.3. Distribution of Drowsiness across Drivers and the Drive

Unlike blood alcohol level and the associated impairment, drowsiness varies considerably across drivers and over the 35-minute drive used in this study. Figure 6 shows the ratings of sleepiness drivers made after they completed each drive using the retrospective sleepiness scale (RSS). Each line represents the ratings of a single driver. The ratings generally increase over the drive. However, these ratings fluctuate considerably from event to event, with uneventful scenario events, such as the straight rural segment, leading to higher ratings of sleepiness. The ratings generally reflect the drowsiness condition, with drivers in the late night condition tending to report higher levels of sleepiness; however, the distribution of reported sleepiness varies considerably with some drivers in the late night condition reporting lower levels of sleepiness compared to those in the daytime condition. Some drivers in the late night condition are quite alert and some in the daytime condition are quite drowsy. This pattern of impairment contrasts with that of alcohol, where BAC level is well-controlled across conditions—no drivers in the zero BAC condition were impaired by alcohol—and the BAC level was relatively constant across the drive. Assuming that BAC level reflects impairment due to alcohol, alcohol-impairment is controlled and constant across the drive. In contrast, Figure 6 shows that the drowsiness conditions induced substantial drowsiness, but that drowsiness varies considerably between drivers, within conditions, and across the drive.

Overall, there were 623 verified lane departures during the drives with 202 being classified as drowsy lane departures. The drowsy departures represented 22 percent of the daytime departures, 14 percent of the early night departures, and 51 percent of the late night departures. Figure 7 shows the frequency of drowsiness-related lane departures, with each line representing data from a single driver. The distribution of these lane departures across the conditions, drive, and drivers shares important features with the ratings of sleepiness. Like the high ratings of sleepiness, more drowsy-related lane departures occurred later in the drive, during long, uneventful segments such as the straight rural and dark rural segments. These peaks likely represent the demands of the roadway (poorly lit and relatively narrow lanes) as well as the association with higher levels of drowsiness. The frequency of lane departures varied considerably across drivers and scenario events with some drivers frequently departing their lane and others departing their lane very infrequently if at all. Similarly, during some scenario events, such as those early in the drive, drivers never departed their lane.

The pattern of drowsiness-related impairment reflected in Figure 7 has several important implications for algorithm development and evaluation, as well as for drowsiness countermeasures. Extreme levels of drowsiness and associated lane departures occur even with seemingly well-rested drivers during the daytime. Unlike alcohol (as suggested by BAC), drowsiness and its effect on lane keeping varies considerably over a drive and across drivers, making the definition of impairment challenging: impairment might not exist for a given driver within a particular scenario event even though the drowsiness condition was designed to induce impairment. Likewise, an otherwise alert driver might experience a period of extreme drowsiness; but when averaged over a drive, the mean level of drowsiness might suggest the driver was safely alert. This makes it less likely that algorithms, such as those used to detect alcohol impairment, will be able to combine event-based (medium range) information to estimate impairment over the drive.

5.4. Detecting Drowsiness with Algorithms Designed for Alcohol Impairment and Distraction

The challenge of detecting drowsiness associated with differences between drivers across the three drowsiness conditions (daytime, early night, and late night) is reflected in the relatively poor detection performance summarized in Table 11. In this table, the algorithms were assessed according to how well they differentiated the day drive from the late night drive using the metrics of AUC, PPP, and accuracy described in Section 5.2. Each algorithm was applied on a long range timescale in which classification instances were accumulated throughout the entire drive.

Not surprisingly algorithms developed to detect distraction failed to detect drowsiness— the AUC of .50 indicates the algorithm performed no better than chance. Surprisingly, algorithms designed to detect drowsiness, such as PERCLOS and those based on EEG measures also performed no better than chance. Poor performance of the algorithms reflects, in part, the drivers in the late night condition who rated themselves as alert and drivers in the daytime condition as very sleepy.

Table 17 shows algorithm performance in detecting drowsiness, as defined by drivers' ratings of sleepiness using the SSS after they completed the drive. Drowsiness is indicated by post SSS of 5 or greater and alertness by post SSS of 3 or less. In this table, the algorithms were assessed according to how well they differentiated between drivers with a rated sleepiness score of 3 or less and those with a score of 5 or greater. Surprisingly, all algorithms performed poorly with only the PERCLOS algorithm having a confidence interval that did not include.50. The mean AUC for the PERCLOS algorithm was only.61, meaning that if the driver was drowsy the algorithm would only have a 61-percent chance of correctly detecting the drowsiness.

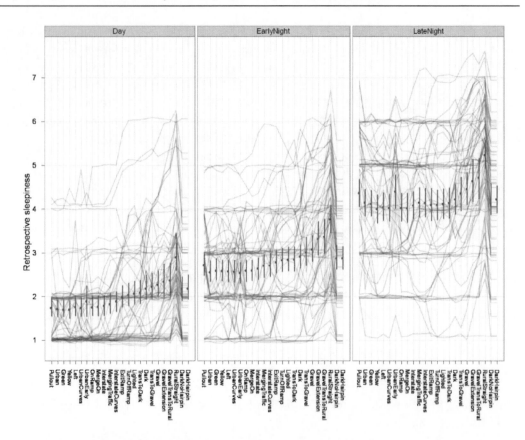

Figure 6. Retrospective sleepiness ratings across the drive. Each line represents a single driver and each point represents the mean with a 95-percent confidence interval.

Table 16. Impairment detection algorithm performance based on drowsiness conditions with 95 percent confidence intervals

Label	Algorithm	AUC	PPP	Accuracy
MDD	Multi-Distraction Detection	.50 (.37-.57)	.52 (.50-.56)	.53 (.49-.53)
EEG	EEG	.54 (.43-.62)	.52 (.50-.53)	.53 (.51-.55)
PC	Perclos	.58 (.49-.67)	.65 (.57-.69)	.61 (.55-.61)
PC+	Perclos+	.51 (.41-.60)	.7 (.54-.80)	.55 (.51-.56)
SB	Steering-Based	.55 (.46-.63)	.55 (.54-.57)	.55 (.55-.56)
BN	Bayes network	.46 (.36-.57)	.50 (.38-.67)	.52 (.51-.53)

Table 17. Impairment detection algorithm performance based on post-drive sleepiness ratings with 95 percent confidence intervals

Label	Algorithm	AUC	PPP	Accuracy
MDD	Multi- Distraction Detection	.51 (.45-.61)	.59 (.55-.62)	.55 (.53-.55)
EEG	EEG	.58 (.48-.65)	.54 (.53-.55)	.59 (.56-.61)
PC	Perclos	.63 (.53-.70)	.60 (.59-.60)	.59 (.55-.61)
PC+	Perclos+	.53 (.43-.60)	.59 (.58-.60)	.54 (.53-.59)
SB	Steering-Based	.55 (.48-.62)	.59 (.58-.59)	.56 (.54-.59)
BN	Bayes network	.45 (.38-.57)	.48 (.45-.51)	.49 (.47-.51)

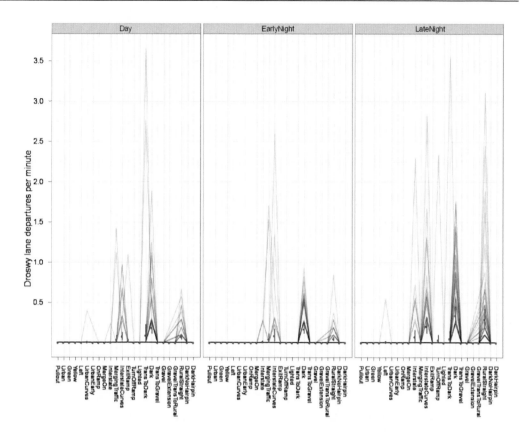

Figure 7. Frequency of drowsiness-related lane departures across the drive.

In both tables the algorithm developed to detect distraction (MDD) performed very poorly. Similarly, the Bayes network trained to detect alcohol impairment also performed very poorly, and algorithms developed to detect drowsiness performed almost as poorly. Overall, these results show that algorithms developed to detect other impairments will not necessarily detect overall drowsiness as determined by SSS rating.

To assess whether algorithms developed to detect alcohol impairment perform better when they are trained to detect drowsiness, the most sensitive algorithm from the IMPACT study—a boosted decision tree using data summarized for each event— was applied to detect drowsiness. Not all measures from IMPACT that were used to train the alcohol algorithm were used in this study, so the original DT algorithm was not used. However, a direct comparison was done with a similar Bayes network algorithm; and the alcohol-trained version did not perform well on drowsiness data (see Table 16 and Table 17). A best-case analysis would consider a DT trained on drowsiness data; and this analysis is presented and showed relatively poor performance. To further tune the DT to detect drowsiness PERCLOS was added to enhance performance.

Once again, post-SSS Ratings were used to classify true drowsiness, and a long-range timescale was used. Figure 8 shows receiver operator curves (ROC) that describe the performance of the algorithms. Comparing the upper panels shows that adding driving performance variables to PERCLOS increases its sensitivity substantially. The graphs in the lower panel show that the driving performance variables and variables that describe the driving context can also be used to detect drowsiness, but less well than PERCLOS. Figure 9 shows the driving performance variables that are most indicative of drowsiness, with lateral and longitudinal acceleration (Ax_max and Ay_max), as well as normalized speed (spn_avg) and lane position (lp_avg) exerting a particularly strong influence. These results show that when trained on data from drowsy drivers the boosted decision tree algorithm can successfully detect drowsiness.

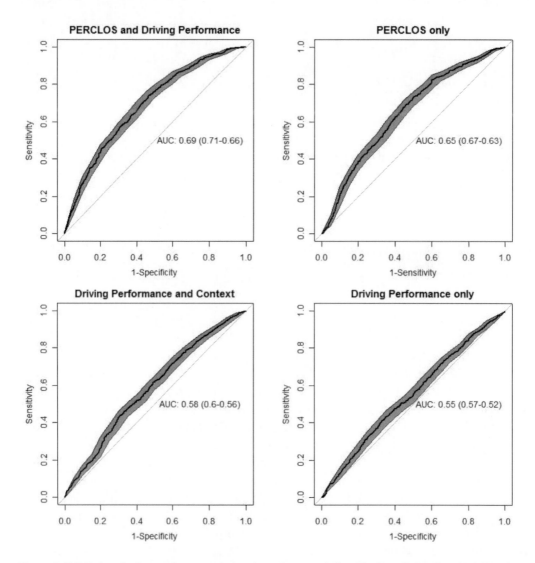

Figure 8. ROC plots for boosted trees to detect drowsiness as defined by Post Drive Stanford Sleepiness Score (5 or greater for drowsy, 3 or less for alert). The upper right ROC uses only PERCLOS, the upper left uses PERCLOS and driving performance and driving context variables. The lower left ROC uses only driving performance and driving context, and the lower right ROC uses only driving performance variables.

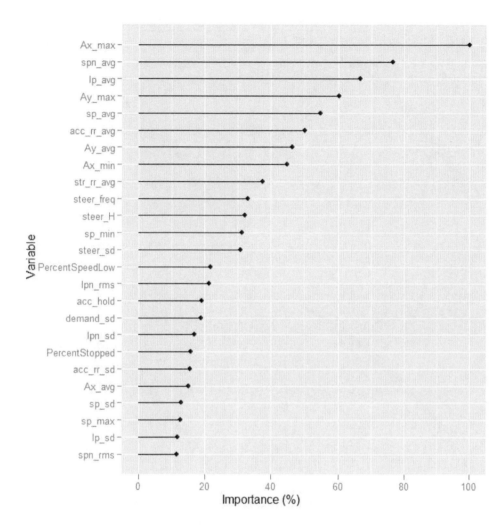

Figure 9. Relative importance of variables predicting drowsiness defined by the post- drive Stanford Sleepiness Score.

If drowsiness is defined by retrospective sleepiness (RSS) ratings rather than post-drive SSS ratings a slightly different picture emerges. Figure 10 shows that boosted trees, detecting event-level measures of sleepiness, perform better than algorithms predicting drowsiness based on the post-drive Stanford Sleepiness Score. Importantly, the algorithms using the driving performance measures perform comparably to PERCLOS. Because sleepiness varied considerably over the drive, it is not surprising that algorithms predicting rated drowsiness for each scenario event performed better than those predicting drowsiness at the end of the drive.

The upper right ROC uses only PERCLOS, the upper left uses PERCLOS and driving performance and driving context variables. The lower left ROC uses only driving performance and driving context, and the lower right ROC uses only driving performance variables.

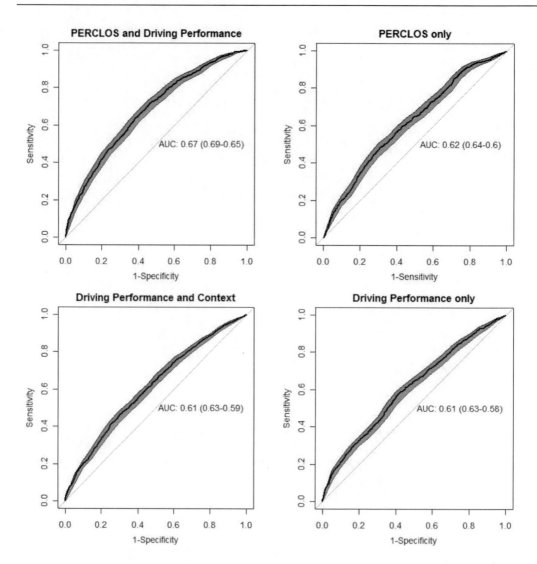

Figure 10. ROC plots for boosted trees to detect drowsiness defined by Retrospective Stanford Sleepiness Score (5 or greater for drowsy, 3 or less for alert).

5.5. Discriminating between Drowsiness and Alcohol Impairment

To more directly assess why algorithms designed to detect alcohol impairment perform poorly in detecting drowsiness, algorithms using long timescales were created using the alcohol data and the drowsiness data. This approach can also evaluate the ability of an algorithm to differentiate between two types of impairment. For this analysis, a Bayes network was selected for investigation. Bayes nets and decision trees are comparable types of machine learning approaches and there is presently no motivation to prefer one over the other. A set of algorithms that range in type is instead provided. Many measures were considered for inclusion in a Bayes network (BN) algorithm. Table 18 summarizes the variables considered; however the classifications were not sensitive to the majority of them.

Table 18. Measures considered for inclusion in the Bayes network

AvgLP > 2	SRR small > 2.2	PERCLOS > 20	Ampd2theta < 50
AvgLP > 1.1	AECS > 1.75	PERCLOS > 5	Ampd2theta < 80
SDLP > 1.3	AECS > 1.2	PERCLOS+ = 3	PRC 17s > 90
SDLP > 1	TLC < 6.5	PERCLOS+ = 2	PRC 60s > 90
SRR > 0.1	TLC < 7.5	Outside% > 50	EEG DCAT = 1
SRR large > 0.03	TLC < 8	Wtflat0 > 300	SpdNorm > 5
SRR small > 2.8	PERCLOS > 40	Wtflat0 > 200	Ax > 0.005

The Bayes network algorithms were developed by computing each measure over a one-minute moving window (except for PERCLOS which is traditionally computed over a three-minute moving window). Threshold values were selected for each measure and each exceedance of the threshold was marked for the entire drive. The rate of exceedance events in a moving six minute window was computed for each measure, and metrics were applied to the rate variable including: average, median, inter-quartile range, 90th percentile, maximum. Additionally, the percentage of time during the drive that the threshold was exceeded was included as a metric. This analysis used the long range timescale that spanned the entire drive.

Estimates of the threshold values were obtained by examining ROC plots for each metric when applied to the data set composed of lane departures associated with drowsiness that were generated through a video review. Those metrics with the highest AUCs were selected for inclusion in the Bayes network.

When this method was applied to the alcohol data, only two measures emerged as indicative of alcohol impairment: small steering reversal rate (SRR small > 2.2) and percent road center gaze (PRC 17s > 90). A binary classification of BAC levels was used that included both .05 and .1 BACs as indicating alcohol-impaired drivers. The average and percent metrics for the first, along with the percent metric for the second measure were used to train the model. Although various depths of graph were tried, a one-level network, also known as a naïve Bayes model, performed the best. ROC performance with 95 percent confidence intervals created with the bootstrap method (Efron & Gong, 1983) is graphed in Figure 11. Point wise confidence intervals are shown by the light colored lines; and the range of AUC values is included in the figure.

The drowsiness dataset was also amenable to this approach. The entire pool of drivers was considered rather than being restricted to verifiably awake subjects, as in the lane departure dataset. Drowsiness was selected as a binary classification, where drowsiness was defined as drives with pre and post SSS scores greater than three; and the alertness was defined as drives with pre and post SSS scores of 3 or less. Drives in which the pre and post SSS scores straddled the threshold were eliminated from the training and test set.

After examining ROC plots for all the measures using the lane departure dataset and the above classification of drowsiness, four measures were included in the drowsiness Bayes network: standard deviation of lane position (SDLP > 1), average eye closure speed (AECS > 1.2), and time to lane crossing (TLC < 6.5, TLC < 7.5), where the average, maximum, maximum, and percentage metrics were applied respectively. The model and ROC performance curve are shown in Figure 12.

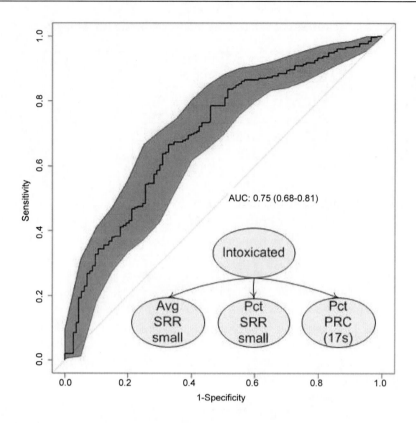

Figure 11. Performance of the Bayes network for detecting alcohol impairment.

On the surface, it would seem that drowsiness and intoxication could be differentiated because a distinct set of measures was used to detect the two impairments. If a measure was used in both models, it would have been because both impairments influenced it. Because alcohol and drowsiness influence driver performance differently, the distinct set of measures suggests some degree of differentiation is possible.

Selecting data for an algorithm to distinguish alcohol impairment and drowsiness presented a challenge. The data for alcohol and drowsiness could not be combined because the thresholding operation was sensitive to minor bias differences in the measures between the two studies. These differences may have been due to small changes in the simulator hardware, software, or protocol between studies. Focusing on the alcohol data exclusively, there were only four drives where the driver was drowsy (post SSS > 3) but not intoxicated, so it was not possible to compare pure drowsiness with intoxication. Instead, a binary class was defined with intoxication and drowsiness as one level, and intoxication but no drowsiness as the other.

The measures that this algorithm used to discriminate between alcohol impairment and drowsiness were a combination of measures used in the previous two models: SRR small > 2.2, SDLP > 1.3, and TLC < 6.5, with average and percentage metrics applied to the first, percentage to the second, and both maximum and percentage applied to the last measure. The model and ROC curve are shown in Figure 13 demonstrate that the effects of alcohol impairment and drowsiness can be distinguished.

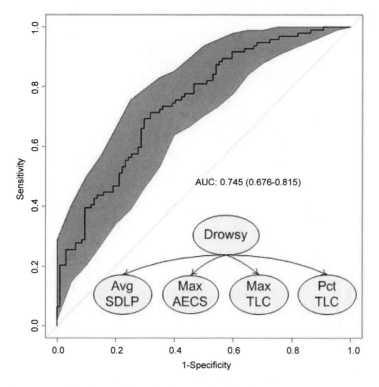

Figure 12. Performance of the Bayes network for detecting drowsiness.

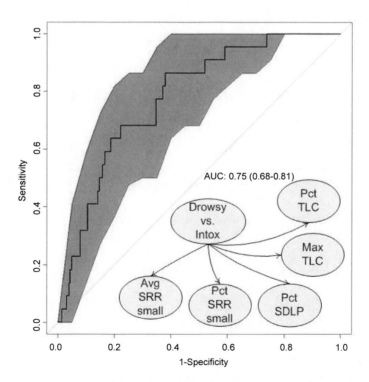

Figure 13. Performance of a Bayes network to differentiate drowsiness combined with alcohol impairment from just alcohol impairment.

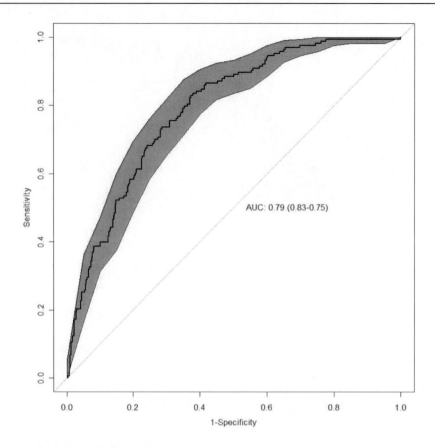

Figure 14. Timeliness using time-to-lane crossing (TLC).

5.6. Detecting Drowsiness Associated with Lane Departures

Given the variability of drowsiness across conditions, drivers, and scenario events across the drive it is not surprising that algorithms detecting impairment defined by the drowsiness condition performed poorly. The transient nature of drowsiness suggests that algorithms that detect impairment associated with driving mishaps, such as lane departures, might be substantially more sensitive.

To assess this possibility, real-time algorithms were developed using short-range timescale continuous data, with a focus on data surrounding lane departures. The continuous data consists of driver and vehicle data recorded at 60 Hz for the entire drive. Each record of these datasets was coded as alert or drowsy according to three definitions: the drowsiness condition (day, early night, late night), a linear combination of PVT, pre- post and retrospective SSS, and the presence or absence of a lane departure. The details of defining truly drowsy lane departures and corresponding truly alert data points are described in Appendix W.

Ten-fold cross validation was used to assess each algorithm, producing a measure of accuracy, PPP, AUC, timeliness and corresponding confidence interval for each algorithm.

Timeliness is defined by the AUC of the ROC curve measured at six seconds before the lane departure. ROC curves summarize the performance graphically.

Time-to-lane-crossing (TLC) is predictive of drowsy lane departures. Although the effectiveness of the classification at the point of departure is trivial and uninteresting because TLC is always equal to zero at this point, the ability of TLC to indicate drowsiness six seconds before a lane departure is very important. TLC is measured here as a moving average over a 60-second window. ROC performance of TLC is shown in Figure 14 below. An AUC of 0.79 of this algorithm shows that the TLC algorithm can identify almost 80 percent of drowsiness-related lane departures before they occur.

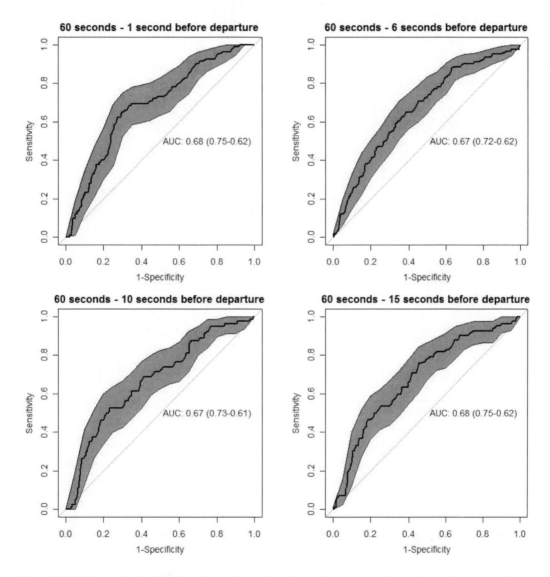

Figure 15. ROC curves for detecting drowsiness-related lane departure, using only continuous steering data with a moving window of 60 seconds.

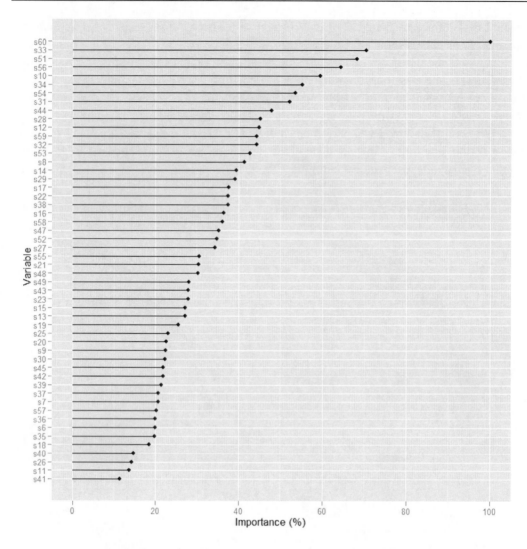

Figure 16. Variable Importance plot for 6 seconds prior classifier. Note that variables are labeled so that "s8" is the steering wheel angle 8 seconds prior to departure.

Steering behavior can also detect drowsiness in advance of lane departures. Figure 15 shows that a relatively simple random forest algorithm (Breiman, 2001) that aggregates the steering wheel position over the previous 60 seconds detects drowsiness substantially better than chance, although not as well as the TLC algorithm. This detection performance is quite timely, detecting drowsiness even 15 seconds before the lane departure. Figure 16 shows the importance of steering wheel position information in detecting drowsiness. Interestingly, the position of the steering wheel at 60, 33, 51, and 56 seconds before the prediction are the most important in detecting drowsiness, showing that steering behavior from across the entire 60-second window preceding a lane departure is useful in predicting lane departures.

The promising performance of both the random forest applied to steering wheel position and the moving average of the TLC contrast with poor performance of PERCLOS. Figure 17 shows that PERCLOS performs only slightly above chance and markedly worse than either the TLC or steering wheel position algorithms. The accuracy of the steering models could likely be improved through data processing and filtering, as well as by combining TLC and steering wheel position information. PERCLOS might provide a useful complement to the steering and lane position algorithms because PERCLOS performs well in the ROC region associated with high specificity, where the algorithm using steering wheel movements performs relatively poorly.

5.7. Conclusions and Implications

The development and evaluation of algorithms to detect drowsiness described in this section provide answers to the four questions that motivated the study.

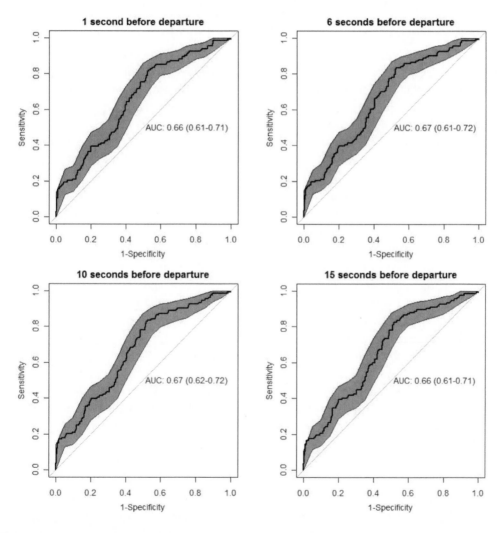

Figure 17. ROC curves for predictive models for drowsiness related lane departure, using only PERCLOS with a moving window of 60 seconds.

Can algorithms designed to detect alcohol impairment and distraction also detect drowsiness? Algorithms developed to detect distraction and alcohol-impaired driving did not detect drowsiness reliably.

Can algorithms designed to detect alcohol impairment be generalized to work well for both alcohol and drowsiness? Algorithms, such as the boosted decision tree that successfully detected alcohol-impaired driving could be generalized to detect drowsiness when trained on drowsy-driver data.

Can algorithms distinguish between alcohol and drowsiness-related impairment? A Bayes network algorithm successfully differentiated alcohol-impaired drivers from drivers who were both drowsy and alcohol-impaired.

Do real-time algorithms perform better in detecting drowsiness in advance of a drowsiness-related mishap? Real-time algorithms, based on lane-keeping and steering behavior, successfully detected drowsiness six seconds before the lane departure. This contrasts with particularly poor performance of PERCLOS in detecting impending lane departures.

Beyond these specific questions, the results of developing and evaluating algorithms to detect drowsiness support important conclusion relative to impairment detection and countermeasure development. The results reinforce earlier findings regarding the qualitative differences between impairments, such as alcohol and distraction. Impairment due to drowsiness and alcohol affects drivers differently, with drowsiness being somewhat transient and alcohol being more persistent, assuming that alcohol impairment is associated with BAC level. The transient nature of drowsiness makes accurate detection of drowsiness at the level of a drive somewhat more difficult than with alcohol impairment.

Beyond the relatively transient nature of drowsiness, the variables most sensitive to detecting each impairment are different. This difference demonstrates the need for separate algorithms to detect the two impairments. Moreover, it was possible to discriminate drowsy intoxicated drivers from non-drowsy intoxicated drivers, showing that the symptoms of intoxication do not necessarily mask those of drowsiness. Ultimately, the results are favorable in regards to the possibility of detecting both drowsiness and intoxication using two independent Bayes network algorithms and discriminating between the two.

Algorithms based on easily accessible measures of steering and lane position performed as well or better than algorithms, such as PERCLOS, that use expensive eye tracking or brain activity sensors. Combining other driving performance measures with PERCLOS leads to substantially better drowsiness detection compared to PERCLOS alone.

Algorithms to detect drowsiness-related lane departures performed very well, providing accurate indications of impending lane departures 6 to 15 seconds before the departure. These algorithms used simple measures of lane keeping and steering behavior. In contrast, PERCLOS performed particularly poorly as a real-time algorithm and depends on a complex sensor to track eye closure. One reason for the poor performance of PERCLOS might be attributed to poor quality eye tracking data and not to the algorithm itself. Such sensitivity to sensor quality represents an important consideration in algorithm design. Accurate measures of lane position are likely to become accessible as lane departure warning systems become more common, and steering behavior can be measured accurately with inexpensive sensors. Naturalistic driving data would better characterize sensor performance because impairment detection depends not only on the algorithm performance, but also on the sensor performance. Alternatively, more accurate sensor models for lane tracking cameras can be added to the simulation. Then the signal- to-noise level can be adjusted and the sensitivity of TLC to sensor accuracy evaluated.

Generally, these results demonstrate the utility in considering indicators of drowsiness beyond PERCLOS in creating real-time algorithms to detect drowsiness.

Although drowsiness produces acute impairment associated lane departures, drowsiness is also revealed with data over a timescale of several minutes. Such long-term drowsiness is revealed by standard deviation of lane position (SDLP), eye closure rate (AECS), and time to lane crossing (TLC). The last measure reinforces the selection of lane departures as an appropriate event to study in relation to impairment. Interestingly, different measures indicate alcohol impairment: small steering reversal rate (SRR) and percent road center (PRC) measure of gaze concentration within a 17-second window. These results suggest that the algorithms to detect long-term drowsiness might be paired with real-time algorithms to improve their performance. Even more broadly, the strong effect of time of day, time spent driving, and even diagnosis of sleep apnea, could further augment the long-term indicator of drowsiness. If such a long-term algorithm indicates the driver is drowsy then the criteria used by the real-time algorithm could be adjusted so that more of the imminent drowsiness-related lane departures are detected before the driver departs the lane.

The success of drowsiness detection algorithms that use low-cost measurements, such as steering inputs, suggests substantial value in further exploration of how such simple sensors can identify impairment. A plan for this exploration would consist of three primary approaches:

1. Investigate the features of the random forest algorithm to understand the features that underlie its success.
2. Apply techniques for impairment detection from time series data including: distribution parameters (mean, standard deviation, kurtosis, etc.), system identification techniques, time-frequency analysis (e.g., Fourier and wavelet analysis), and symbolic aggregate approximation (SAX) time series analysis.
3. Develop hierarchical, variable-time-window algorithms. Such algorithms integrate information from a long time scale, such as the time of day, with information from a short time scale, such as the previous minute of steering behavior.

These approaches support a deeper understanding of the data that can detect and discriminate impairments using simple sensors, such as steering wheel instrumentation. The hierarchal algorithm will indicate how best to combine such data to improve detection and discrimination performance.

REFERENCES

Adan, A. & Almirall, H. (1991). Horne and Osterberg Morningness Eveningness Questionnaire - a Reduced Scale. *Personality and Individual Differences 12*(3):241-253.

Advanced Brain Monitoring (Designer). (2011). *B-Alert x-10.* (Print Drawing). Retrieved from www.b-alert.com/pdf/x10.pdf.

Ambulatory Monitoring Incorporated. (Designer). (2009). *Motionlogger Acigraph*. Retrieved from www.ambulatory-monitoring.com/motionlogger.html.

Arnedt, J. T., Wilde, G. J. S., Munt, P. W. & MacLean, A. W. (2001). How do prolonged wakefulness and alcohol compare in the decrements they produce on a simulated driving task? *Accident Analysis and Prevention, 33*, 337−344.

AWAKE. (2010). AWAKE - System for Effective Assessment of Driver Vigilance and Warning According to Traffic Risk Estimation. www.awake-eu.org/.

Barr, L., Howarth, H., Popkin, S. & Carroll, R. J. (2003). *A review and evaluation of emerging driver fatigue detection measures and technologies*. Cambridge, MA: Volpe National Transportation System Center.

Beirness, D. J., Simpson, H. M. & Desmond, K. (2005). *The Road Safety Monitor 2004: Drowsy Driving*. Ottawa, ON: Traffic Injury Research Foundation.

Breiman, L. (2001), Random Forests, *Machine Learning, 45*, 5-32.

Brown, T., Dow, B., Trask, D., Dyken, E. & Salisbury, S. E. (2009). Driving Performance and Obstructive Sleep Apnea: A Preliminary Look at the Manifestations of Impairment. *Transportation Research Record: Journal of the Transportation Research Board, 2096*, 33-40.

Dawson, Drew. & Kathryn, Reid. (1997). Fatigue, alcohol and performance impairment. *Nature, 388* (6639): 235.

De Valck, E. & Cluydts, R. (2001). Slow-release caffeine as a countermeasure to driver sleepiness induced by partial sleep deprivation. *Journal of Sleep Research, 10*, 203−209.

Dinges, D. F., Mallis, M. M., Maislin, G. & Powell IV, J. W. (1998). *Evaluation of Techniques for Ocular Measurement as an Index of Fatigue and as the Basis for Alertness Management*. (Report No. DOT HS 808 762). Washington, DC: National Highway Traffic Safety Administration. Available at http://ntl.bts.gov/lib/jpodocs/edlbrow/7d01!.pdfv

Efron, B. & Gong, G. (1983). A leisurely look at the bootstrap, the jackknife and cross-validation. *American Statistician, 37*(1), 36-48.

Fairclough, S. & Graham, R. (1999). Impairment of Driving Performance Caused by Sleep Deprivation or Alcohol: A Comparative Study. *Human Factors, 41*(1), 118-128.

Hanowski, R. J., Bowman, D. S., Alden, A., Wierwille, W. W. & Carroll, R. (2008a). PERCLOS+: Moving Beyond Aingle-Metric Drowsiness Monitors. Society of Automotive Engineers Commercial Vehicle Engineering Conference.

Hanowski, R. J., Bowman, D. S., Wierwille, W. W., Alden, A. & Carroll, R. (2008b). PERCLOS+: Development of a robust field measure of driver drowsiness. Proceedings of the 15th World Congress on Intelligent Transport Systems, New York, November 2008.

Hoddes, E., Zarcone, V., Smythe, H., Phillips, R. & Dement, W. C. (1973). Quantification of Sleepiness: A New Approach. *Psychophysiology, 10*(4), 431-436.

Intoximeters Inc. *Alcosensor iv*. (2009). [Web Document]. Retrieved from www.intox.com/p-559-alco-sensor-iv.aspx.

Ji, Q., Zhu, Z. & Lan, P. (2004). Real-Time Nonintrusive Monitoring and Prediction of Driver Fatigue. *IEEE Transactions on Vehicular Technology , 53*(4).

Klauer, S. G., Dingus, T. A., Neale, V. L., Sudweeks, J. D. & Ramsey, D. J. (2006). *The Impact of Driver Inattention on Near-Crash/Crash Risk: An Analysis Using the 100-Car Naturalistic Driving Study Data*. (Report No. DOT HS 810 594). Washington, DC: National Highway Traffic Safety Administration.

Kloss, J. D., Szuba, M. P. & Dinges, D. F. (2002). Sleep loss and sleepiness: physiological and neurobehavioral effects. Chapter 130. In: L. Davis, D. Charney, J. T. Coyl, & C. Nemeroff, eds. *Neuropsychopharmacology: The Fifth Generation of Progress.* Philadelphia: Lippincott Williams & Wilkins.

King, D. J., Mumford D . K. & Siegmund, G. P. (1998). *An Algorithm for Detecting Heavy-Truck Driver Fatigue from Steering wheel Motion.* Paper Number 98-S4- O-10, 873-882. Richmond, BC: MacInnis Engineering Associates Ltd.

Knipling, R. R. & Shelton, T. T. (1999). Problem Size Assessment: Large Truck Crashes Related Primarily to Driver Fatigue. *Proceedings of the Second International Large Truck Safety Symposium,* EO1-2510-002-00, University of Tennessee Transportation Center, Knoxville, Tennessee, October 1999.

Lee, J., Fiorentino, D., Reyes, M., Brown, T., Ahmad, O., Fell, J. & DuFour, R. (2010). *Assessing the Feasibility of Vehicle-Based Sensors to Detect Alcohol Impairment.* (Report No. DOT HS 811 358). Washington, DC: National Highway Traffic Safety Administration.

Lee, J. D., Moeckli, J., Brown, T. L., Roberts, S. C., Schwarz, C., Yekshatyan, L. & Davis, C. (In review). *Distraction detection and mitigation through driver feedback.* Washington, DC: National Highway Traffic Safety Administration.

Loh, S., Lamond, N., Dorrian, J. Roach, G. & Dawson, D. (2004). The validity of psychomotor vigilance tasks of less than 10-minute duration. *Behavior Research Methods, Instruments, & Computers, 36*(2), 339-346.

Liu, C. C., Hosking, S. G. & Lenne, M. G. (2009). Predicting driver drowsiness using vehicle measures: Recent insights and future challenges. *Journal of Safety Research,* doi:10.1016/j.jsr.2009.04.005.

Mackie, R. R. & Miller, J. C. (1978). *Effects of hours of service, regularity of schedules and cargo loading on truck and bus driver fatigue.* (Technical Report 1765-F). Goleta, CA: Human Factors Research, Inc.

Matthews, G. & Desmond, P. A. (2002). Task-induced fatigue states and simulated driving performance. *Quarterly Journal of Experimental Psychology: Human Experimental Psychology, 55A,* 659–686.

Mattsson, K. (2007). *In-Vehicle Prediction of Truck Driver Sleepiness: Lane Position Variables.* Master's Thesis. Luea University of Technology.

Maycock, G. (1997). Sleepiness and driving: the experience of U.K. car drivers. *Acid. Anal. Prev. 29,* 453-462.

National Highway Traffic Safety Administration. (2008). *National Motor Vehicle Crash Causation Survey. Report to Congress.* (Report No. DOT HS 811 059). Washington, DC: Author.

NHTSA. (2011, March). NHTSA Traffic Safety Facts on Drowsy Driving. Traffic Safety Facts (Report No. DOT HS 811 44). Washington, DC: Author.

National Sleep Foundation. (2009). *2009 Sleep in America Poll: Summary of Findings.* Arlington, VA: Author

Nissan. (2011). *Drunk-driving Prevention Concept Car.* Retrieved March 24, 2011, from www.nissan-global.com/EN/TECHNOLOGY/OVERVIEW/dpcc.html.

Otmani, S., Pebayle, T., Roge, J. & Muzet, A. (2005). Effect of driving duration and partial sleep deprivation on subsequent alertness and performance of car drivers. *Physiology & Behavior, 84,* 715−724.

Philip, P., Sagaspe, P., Taillard, J., Valtat, C., Moore, N., Akerstedt, T. & Bioulac, B. (2005) Fatigue, sleepiness and performance in simulated versus real driving conditions. *Sleep*, *28*,1511– 6.

Platt, F. N. (1963). A new method of measuring the effects of continued driving performance. *Highway Research Record*, *25*, 33-57.

Regan, M. A., Lee, J. D. & Young, K. L., (2009*). Driver distraction: Theory, effects, and mitigation*. New York: Taylor and Francis.

Ridder, T., Maynard, J., Abbink, R., Johnson, R., Hull, E., Meigs, A., et al. (2008). *Noninvasive Determination of Alcohol in Tissue* (2006.01 ed.). United States Patent No. US7403804 B2. Tru Touch Technologies, Inc.

Royal, D. (2003). *National Survey of Distracted and Drowsy Driving Attitudes and Behavior:2002 Volume I: Findings*. (Report No. DOT HS 809 566). Washington, DC: National Highway Traffic Safety Administration.

Safford, R. R. & Rockwell, T. H. (1967). Performance decrement in twenty-four hour driving. *Highway Research Record*, *163*, 68-79.

Thiffault, P. & Bergeron, J. (2003). Monotony of Road Environment and Driver Fatigue: A Simulator Study, *Accid. Anal. Prev.*, *35*, 381.

Tijerina, L., Gleckler, M., Stolzfus, D., Johnston, S., Goodman, M. J. & Wierwille, W. W. (1999, March). A Preliminary Assessment of Algorithms for Drowsy and Inattentive Driver Detection on the Road. Washington, DC: National Highway Traffic Safety Administration.

Webster, G. & Gabler, H. (2007). Feasibility of Transdermal Ethanol Sensing for the Detection of Intoxicated Drivers. Melbourne, Australia: Association for the Advancement of Automotive Medicine.

Wierwille, W. W., Ellsworth, L. A., Wreggit, S.S., Fairbanks, R. J. & Kim, C. L. (1994). *Research on vehicle-based driver status/performance monitoring: development, validation, and refinement of algorithms for detection of driver drowsiness*. (Report No. DOT HS 808 247). Washington, DC: National Highway Traffic Safety Administration.

Wierwille, W. W., Lewin, M. G. & Fairbanks, R. J. (1996). *Final Report: Research on vehicle-based driver status/performance monitoring, Part I*. (Report No. DOT HS 808 638). Washington, DC: National Highway Traffic Safety Administration.

Wierwille, W. W., Lewin, M. G. & Fairbanks, R. J. (1996). *Final Report: Research on vehicle-based driver status/performance monitoring, Part II*. (Report No. DOT HS 808 638). Washington, DC: National Highway Traffic Safety Administration.

Williamson, A. M. & Feyer, A. M. (October 2000). Moderate sleep deprivation produces impairments in cognitive and motor performance equivalent to legally prescribed levels of alcohol intoxication. *Occup Environ Med*, *57* (10), 649–55.

Wilson, T. & Greensmith, J. (1983) Multivariate analysis of the relationship between drivometer variables and drivers' accident, sex, and exposure status. *Human Factors*, *25*, 303-312.

Wilkinson, R. T. & Houghton, D. (1982). Field test of arousal: A portable reaction timer with data storage. *Human Factors*, *24*, 487- 493.

End Notes

[1] A total of 103 participants were enrolled.

[2] It should be noted that there was a tradeoff in presenting the CD task between temporary arousal of the driver that might lessen the drowsiness effects, and the ability to compare back to the alcohol data and in the future to begin to examine the interaction between drowsiness and distraction. It was decided that consistency with previous data was more important.

In: Feasibility of Vehicle-Based Sensors
Editor: Meaghan Sadler

ISBN: 978-1-61728-349-9
© 2014 Nova Science Publishers, Inc.

Chapter 2

ASSESSING THE FEASIBILITY OF VEHICLE-BASED SENSORS TO DETECT ALCOHOL IMPAIRMENT[*]

*John D. Lee, Dary Fiorentino, Michelle L. Reyes,
Timothy L. Brown, Omar Ahmad, James Fell, Nic Ward
and Robert Dufour*

ABSTRACT

Despite persistent efforts at the local, state, and federal levels, alcohol-impaired driving crashes still account for 31% of all traffic fatalities. The proportion of fatally injured drivers with blood alcohol concentrations (BAC) greater than or equal to 0.08% has remained at 31-32% for the past ten years. Vehicle-based countermeasures have the potential to address this problem and save thousands of lives each year. Many of these vehicle-based countermeasures depend on developing an algorithm that uses driver performance to assess impairment. The National Advanced Driving Simulator (NADS) was used to collect data needed to develop an algorithm for detecting alcohol impairment. Data collection involved 108 drivers from three age groups (21-34, 38-51, and 55-68 years of age) driving on three types of roadways (urban, freeway, and rural) at three levels of alcohol concentration (0.00%, 0.05%, and 0.10% BAC). The scenarios used for this data collection were selected so that they were both representative of alcohol-impaired driving and sensitive to alcohol impairment. The data from these scenarios supported the development of three algorithms. One algorithm used logistic regression and standard speed and lane-keeping measures; a second used decision trees and a broad range of driving metrics that are grounded in cues NHTSA has suggested police officers use to identify alcohol-impaired drivers; a third used a support vector machines. The results demonstrate the feasibility of a vehicle-based system to detect alcohol impairment based on driver behavior. The algorithms differentiate between drivers with BAC levels at and above and below 0.08%BAC with an accuracy of 73 to 86%, comparable to the standardized field sobriety test. This accuracy can be achieved with approximately eight minutes of driving performance data. Differences between drivers and between roadway

---

[*] This is an edited, reformatted and augmented version of a National Highway Traffic Safety Administration sponsored document, DOT HS 811 358, issued August 2010.

situations have a large influence on algorithm performance, which suggests the algorithms should be tailored to drivers and to road situations.

1. EXECUTIVE SUMMARY

The most notable findings from this study include:

- The National Advanced Driving Simulator (NADS-1) and a low-workload scenario are sensitive to alcohol. Alcohol effects were apparent in a simulation scenario representing a typical drive home at night from an urban bar.
- These effects were evident in drivers' control of vehicle lane position and speed. Standard deviation of lane position and average speed differentiated BAC conditions most precisely.
- The most sensitive indicators of impairment are associated with continuous performance (e.g., lane keeping) rather than discrete events (e.g., response to a traffic signal or use of turn signals).
- The three algorithms detected impairment at and above the legal limit about as well as the Standardized Field Sobriety Test (SFST), with sensitivity increasing with BAC level.
- This project demonstrates the feasibility of a driving-behavior-based approach to detecting alcohol impairment in real time.

Background

Despite persistent efforts at the local, state, and federal levels, alcohol-impaired driving crashes still contribute to approximately 31% of all traffic fatalities. Although regulatory and educational approaches have helped reduce alcohol-related fatalities, other approaches merit investigation. One such approach detects alcohol impairment in real time using the increasingly sophisticated sensor and computational platform that is available on many production vehicles.

It may be possible to detect impairment using driver state (e.g., eye movements), drivers' control inputs (e.g., steering and accelerator movements), and vehicle state (e.g., speed or lane position). Once detected, this information can support interventions that discourage drivers from driving while impaired and prevent alcohol-related crashes. This study assessed how well algorithms could detect impairment in a widely applicable and timely manner.

Objectives

The long-term research objective is to use algorithms that detect impairment as feedback to drivers to discourage or prevent drinking and driving. This report describes how, individually and in combination, driver actions reveal signatures of alcohol impairment, and how well algorithms built on these signatures detect drivers with BAC levels that are over the legal limit. Specific objectives include:

- Understand how driving-related metrics reflect the impairment associated with BACs at 0.05% and 0.10% (currently, the legal limit in the United States is 0.08%)
- Determine how well these metrics apply to different roadway situations and to different drivers (i.e., determine robustness)
- Develop algorithms to detect alcohol-related impairment
- Compare robustness of metrics and algorithms

Method

Data were collected in the National Advanced Driving Simulator (NADS) from 108 moderate to heavy drinkers dosed at placebo (0.00%), 0.05%, and 0.10% BAC on three separate visits. Drivers were divided into equal groups by age (21-34, 38-51, and 35-68) and gender. The participants drove a scenario representative of a drive home from an urban bar: a nighttime trip that involved a stretch of freeway and ended on a rural gravel road. The drives started with an urban segment composed of a two-lane roadway through a city with posted speed limits of 25 to 45 mph with signal-controlled and uncontrolled intersections. An interstate segment followed that consisted of a four-lane divided expressway with a posted speed limit of 70 mph. The drives concluded with a rural segment composed of a two-lane undivided road with curves. A portion of the rural segment was gravel. Drivers' steering, accelerator, and brake inputs, vehicle lane position and speed, the driving context (whether the vehicle was in a urban, interstate, or rural environment), and driver eye and eyelid movements were captured in representative driving situations, with precise control and in great detail.

Results

The objectives were addressed with two broad sets of analyses. The first focused on whether BAC affected performance. The second focused on detection of impairment. These analyses show the simulator and scenario to be sensitive to alcohol, and that algorithms can detect alcohol-related impairment.

Driving performance measures (i.e., mean speed, standard deviation of speed, and standard deviation of lane position) indicated systematic differences between BAC conditions. No statistically reliable effects of age and gender were found for lane deviation, but BAC affected lane deviation. Normalized lane deviation for the entire drive was 46.77 at 0.00% BAC, 49.79 at 0.05% BAC, and 54.31 at 0.10% BAC. Age reliably affected average speed, with average speed increasing with increasing age. BAC also affected average speed, with a higher BAC, in general, leading to lower average speed. Neither age nor BAC reliably affected speed deviation. Surprisingly, gender reliably affected speed deviation, with speed deviation greater for males than for females. These results are notable because the alcohol effects are apparent even though all participants were moderate to heavy drinkers and the driving situation was representative of daily driving, placing relatively low demands on the driver.

Taken together, the results from the impairment analyses indicate that alcohol affected performance, and that the NADS is sensitive to those changes. The next set of analyses focused on whether it is possible to classify BAC status (BAC < 0.08% v. BAC ≥ 0.08%) on the basis of those changes. Three algorithms were developed to predict BAC status based on logistic regression, decision tree modeling, and support vector machines. The algorithms were assessed in terms of sensitivity, robustness, timeliness, and bias.

Sensitivity is the degree to which the algorithm can differentiate between drivers above and below the legal limit. These algorithms show sensitivity comparable to that of the SFST. Accuracy of the algorithms ranged from 73 to 86%. The logistic regression algorithm achieved an accuracy of 74.4% by combining information across the entire drive, achieving maximum sensitivity after approximately 25 minutes of driving. Decision tree and Support Vector Machine algorithms are much more sensitive and timely, identifying impairment in the situations tested with greater precision after only approximately eight minutes of driving. Greatest sensitivity was achieved by the decision tree, which combined driving performance indicators tailored to individual drivers. The most sensitive indicators of impairment were associated with continuous performance (e.g., lane keeping) rather than discrete events (e.g., response to a traffic signal or use of turn signals). Performance of the algorithms showed substantial differences in the degree to which they and their constituent measures provide robust and timely indications of impairment.

Robustness is insensitivity to confounding factors, such as different driving environments, and it applies to many factors affecting algorithm performance. An important element of robustness addressed in this study concerns the dependence of the algorithm on differences between drivers and the driving environment (i.e., urban, freeway, rural). Consistent with previous research, algorithms tailored to individuals outperformed generic algorithms by approximately 13%. Algorithm performance also depends on the driving context: different driving situations provide different measures, and these measures differ in their sensitivity. Current vehicle technology makes it quite feasible to capitalize on the benefits of tailoring algorithms to individuals by comparing a driver's performance to his or her past performance in similar roadway situations.

Timeliness is the speed with which an algorithm is able to detect impairment. Timeliness is a critical consideration for real-time algorithms because some interventions rely on impairment detection well before the end of the drive. Timely impairment detection depends critically on the driving context with some events and variables being more sensitive than others. The most sensitive indicators of impairment involve continuous measures cumulated over time, such as the standard deviation of lane position. In addition, important signatures of alcohol impairment (straddling, weaving, and gaze concentration) are defined by behavior that evolves over a relatively long time horizon, requiring samples of driving behavior that extends over 30 seconds to several minutes. Even with such constraints, sensitivity comparable to the SFST was obtained over approximately eight minutes of driving.

Bias refers to the tendency of the algorithm to favor correctly detecting impairment at the expense of incorrectly identifying impairment when there is none. The fundamental differences in the optimization approaches between decision trees and SVM lead to differences in bias. These complementary differences can be leveraged to minimize false detection of impairment and to maximize detection of impaired drivers. Algorithms can be combined according to the benefits of detecting impairment and the costs of failing to detect impairment, so that one algorithm is used to maximize impairment detection and another is used to avoid false detection. Support Vector Machines show a tendency to outperform decision trees in maximizing detection.

Recommendations and Conclusion

This study demonstrates the feasibility of vehicle-based sensors to detect alcohol-related impairment in real time: sensitivity is comparable to the SFST. This sensitivity is likely a very conservative estimate relative to sensitivity in detecting higher BAC levels. Because 66% of alcohol-related fatalities occur with BAC levels above 0.15% (compared with impaired drivers at 0.08 BAC or greater), the greatest value of a vehicle-based countermeasure may lie in detecting high BAC levels, where algorithms are likely to be very sensitive. These results suggest substantial promise in detecting other impairments, such as drowsiness, distraction, and even age-related cognitive decline.

The ultimate aim of impairment-detection algorithms is to support interventions that guide the driver to safer behavior. The desirability and feasibility of any particular algorithm depends on how it meets the particular needs of an intervention. This study demonstrates that algorithms differ substantially on these dimensions and that design must consider the inevitable tradeoffs. Algorithms become more sensitive, but less timely, as measures are integrated over time. The ultimate feasibility of impairment-detection algorithms depends on matching the performance profile of an algorithm to an intervention.

This project demonstrated the feasibility of a behavior-based approach to detecting alcohol impairment. It also identified many issues that merit further investigation. This study focused on moderate levels of alcohol (0.05% and 0.10% BAC) in people who reported to be moderate to heavy drinkers, but not problem drinkers. Given the assumption that moderate and heavy drinkers show less obvious indications of impaired driving, algorithms that are able to detect impaired driving from this population are likely to be much more effective for people who are light drinkers or for higher BAC levels, whereas the effectiveness of the algorithms for problem drinker is unknown. It would be useful to empirically assess how algorithm sensitivity differs at higher BACs, and for light and chronic drinkers.

More generally, this study identified a huge design space of sensors, measures, signatures and algorithms, algorithm parameter combinations, and meta algorithms. Assuming there are at least 10 sensors, 10 metrics for each sensor, 4 time scales, 10 algorithms, 10 implementations of each algorithm, and 4 meta algorithms, a total of more than 160,000 potential algorithms exist. This project sampled only a small region of that space. A specific challenge facing deployment of the algorithms developed in this study is the reliable measurement of lane position. Current lane- tracking technology is vulnerable to sun glare, adverse weather, and poorly maintained lane markings. Steering wheel position is not subject to these limits and further analysis of its ability to detect alcohol-impaired driving is warranted. More generally, a promising direction for algorithm development is to identify classes of drivers and classes of driving situations, and an important direction for algorithm assessment is to develop metrics that relate to the interventions the algorithm intends to support. These results also suggest that further exploration is warranted, not just for alcohol impairment detection, but also for detecting impairment associated with drowsiness, distraction, and even age-related cognitive decline.

2. INTRODUCTION AND OBJECTIVES

Despite persistent efforts at the local, state, and federal levels, alcohol-impaired driving crashes still contribute to approximately 31% of all traffic fatalities. The proportion of fatally injured drivers with blood alcohol concentrations (BAC) greater than or equal to 0.08% has remained at 31-32% for the past ten years (National Highway Traffic Safety Administration, 2009b). Although regulatory and educational approaches have helped to reduce alcohol-related fatalities, other approaches merit investigation. One such approach concerns countermeasures that capitalize on the increasingly sophisticated sensor and computational platform that is available on many production vehicles. Such vehicle-based countermeasures have the potential to address this problem and save thousands of lives each year.

Vehicle-based countermeasures use sensors that describe drivers' control inputs (e.g., steering wheel and brake pedal movement), vehicle state (e.g., accelerometer and lane position), driving context (e.g., speed zone information and proximity of surrounding vehicles), and driver state (e.g., eye movements and posture). Data from these sensors can be transformed, combined, and processed with a variety of algorithms to develop a detailed description of the driver's response to the roadway. These sensors and algorithms hold promise for identifying a range of driver impairments, including distraction, drowsiness, and even age-related cognitive decline. Alcohol represents a particularly important impairment that might be detected by vehicle-based sensors and algorithms.

This report describes the development and evaluation of algorithms to detect the behavioral signature of alcohol. Such an algorithm is a central element of any vehicle-based countermeasure for alcohol-related crashes. Algorithm development depends on collecting data from impaired and unimpaired drivers. This report describes data collection that involved drivers from three age groups (21-34, 38-51, and 55-68 years of age) driving through representative situations on three types of roadways (urban, freeway, and rural) at three levels of alcohol concentration (0.00%, 0.05%, and 0.10% BAC). The high fidelity of the National Advanced Driving Simulator (NADS) makes these data unique. Drivers' control inputs, vehicle state, driving context, and driver state were captured in representative driving situations, with precise control and in great detail. This report describes how, individually and in combination, these data reveal signatures of alcohol impairment, and how well algorithms built on these signatures detect drivers with BAC levels that are over the legal limit of 0.08%.

The overall objectives of the data collection and analysis efforts were to:

- Understand how driving-related metrics reflect the impairment associated with BAC at 0.05% and 0.10%
- Determine the robustness of these metrics with respect to individual differences such as age and gender, as well as the roadway situation
- Identify signatures of impairment and develop algorithms to detect alcohol-related impairment
- Compare robustness of metrics and algorithms

This document contains six main sections. The first describes the prevalence and consequences of drinking and driving. The second describes the sensors and metrics that are likely to capture a clear behavioral signature of alcohol impairment. The third section describes the criteria and considerations for detecting impairment. The fourth section outlines the experimental methods and experimental design, independent variables, dependent variables, and data collection procedures. The fifth section describes the general effects of three alcohol levels on driver performance. The sixth section describes the characteristics of algorithms used to detect alcohol impairment, their performance, and the validation process.

This report focuses on detecting alcohol impairment, but the issues are common to any vehicle- based system that detects driver impairment. As manufacturers work to make vehicles increasingly aware of the roadway and driver state, such systems will become important contributors to vehicle safety. Consequently, this report provides an example of assessing the limits and capabilities of vehicle-based systems that support a behavior-based estimation of driver impairment.

The long-term research objective is to use algorithms that detect impairment to provide feedback to drivers that will discourage drinking and driving. Ultimately the alcohol-impairment-detection algorithms developed in this study could support a range of vehicle-based interventions to prevent alcohol-related crashes. Such interventions could include limiting drivers' ability to drive dangerously (e.g., lockout distractions or limit speed), providing feedback to impaired drivers that may motivate them to pull over or drive more cautiously, adjusting crash warning systems to provide an earlier warnings, or providing long-term feedback that highlights dangerous driving. This approach to detecting alcohol impairment and associated countermeasures complements in- vehicle technology that prevents an alcohol-impaired driver from starting the car, such as that being pursued in the Driver Alcohol Detection System for Safety (DADSS, www.dadss.org/). This dual-prong approach is consistent with many public health programs, ranging from preventing teen pregnancy to reducing smoking-related illnesses.

Using vehicle-based technology to prevent alcohol-impaired drivers from crashing also begins to address the more general problem of preventing drivers suffering from a broad range of other impairments from crashing. The combined influence of alcohol, distraction, fatigue, and drug- and age-related cognitive impairments represent the predominant cause of crashes. Mitigating these impairments could have profound safety benefits.

3. DRINKING AND DRIVING PREVELANCE

Impaired driving is a significant public health and safety problem in the United States (U.S.). It was estimated by the National Highway Traffic Safety Administration (NHTSA) that impaired driving costs society $51 billion annually (Blincoe, et al., 2002). Although alcohol consumption is legal for U.S. citizens aged 21 or older, it is illegal in all states to drive with a blood alcohol concentration (BAC) of 0.08% or greater. In 2008, there were an estimated 11,773 traffic crash fatalities involving drivers at these illegal BAC limits (BAC=0.08% and above) (National Highway Traffic Safety Administration, 2009b).

3.1. Drinking in America

A report by the National Center for Health Statistics (NCHS) on the health behaviors of adults estimated that 6 of 10 people in the United States were drinkers of alcohol in 1999-2001 (consumed alcohol within the past year), whereas about one in four were total abstainers (Public Health Service, 2004). About 20% of the adults in that survey reported consuming five or more drinks in a day at least once in the past year. A Gallup Poll (Saad, 2003) indicated that the percentage of adult drinkers who consumed one to seven drinks in the past week has increased from 36% in 1992 to 45% in 1997 to 50% in 2001. The proportion of drinkers in that same survey who reported consuming eight or more drinks in the past week increased from 12% in 1992 and 1997 to 18% in 2001. In a separate analysis of Gallup's surveys on alcohol and drinking from 1999 to 2003, 40% of men aged 18-29, 29% of men aged 30-49, 21% of men aged 50-64, and only 16% of men aged 65 and older admitted to drinking more than they should (Carroll, 2003). Carroll also found that women of all ages were much less likely to report drinking more than they should at times: 26% of women aged 18-29; 19% of women aged 30-49; 12% of women aged 50-64; and only 5% of women aged 65 and older. Alcohol consumption represents a prevalent and increasingly common behavior.

Although moderate consumption describes the drinking patterns of many, the prevalence of excessive drinking represents an important concern. Recent studies and surveys have indicated that about 8% of the U.S. population can be classified as alcohol dependent or alcohol abusers (NIAAA, 2004). This translates to about 17.6 million American adults, most of whom drive motor vehicles. Binge drinking (defined as consuming 5 or more drinks in one session) increased in the United States between 1995 and 2001, according to a recent survey (Naimi et al., 2003). Binge-drinking episodes per person per year increased 35% during that same period, with men accounting for 81% of the episodes. Overall, 47% of binge-drinking episodes occurred among otherwise moderate drinkers, and 73% of all binge drinkers were classified as moderate drinkers in that study. Binge drinkers were 14 times more likely to drive impaired than non-binge drinkers. The patterns of moderate and excessive drinking describe a societal context in which drinking is likely to precede or accompany driving, and that the societal trends will require countervailing forces to prevent alcohol-impaired driving from reflecting similar trends.

3.2. Drinking and Driving on U.S. Roads

Our knowledge about the impaired-driving problem in the United States has been augmented by a series of National Roadside Survey (NRS) studies from which we can estimate the prevalence of drinking and driving over time in the contiguous 48 states by randomly selecting drivers from the road and requesting breath samples. The first NRS, sponsored by NHTSA, was conducted in 1973 (Wolfe, 1974). The second NRS was sponsored by the Insurance Institute for Highway Safety (IIHS) in 1986 (Lund & Wolfe, 1991), and the third was jointly funded by IIHS and NHTSA in 1996 (Voas, Wells, Lestina, Williams, & Greene, 1998). The first three surveys (1973, 1986, and 1996) included a brief interview of randomly selected drivers, and a breath sample to measure the BAC. The fourth in this series of national surveys, conducted in 2007, followed the general methodology of the three prior surveys in obtaining BACs for comparison with the earlier surveys, but also incorporated several new features (Lacey, et al., 2009). These included questionnaires on drivers' drug use, interaction with the criminal justice and treatment systems, drug- and alcohol-use disorders, and collecting and analyzing oral fluid and blood to determine the presence of drugs (over-the-counter, prescription, and illegal) other than alcohol.

Figure 1 summarizes and compares the results of the four NRS studies of weekend nighttime drivers. The figure shows that the percentage of drivers in all BAC range categories has decreased in succeeding decades, with the exception of an increase in the percentage of drivers with BACs between 0.050% and 0.079% between 1986 and 1996. However, the overall percentage of positive BAC drivers decreased between 1986 and 1996 (Fell, Tippetts, & Voas, 2009).

Based upon data collected in the 2007 NRS, the best predictor of being a drinking driver on U.S. roads was reported as binge drinking (six or more drinks in a session for males; five or more drinks for females). The data showed that people who reported binge drinking were more likely to have had a positive BAC on the road. Given that binge drinkers were more likely to be impaired drivers, binge drinking was a better predictor than being classified as alcohol dependent or an alcohol abuser, as defined by the *Alcohol Use Disorders Identification Test* (AUDIT) (Babor, de la Fuente, Saunders, & Grant, 1992; Chung, Colby, Barnett, & Monti, 2002; Conley, 2001).

Telephone surveys provide a less direct estimate of drinking and driving behavior, but one that is largely consistent with the roadside surveys. In a telephone survey of more than 6,000 people aged 16 and older in the United States in 2001, 23% reported driving within two hours of drinking alcohol in the past year (Royal, 2003). In that same survey, problem drinkers were estimated to make up 29% of the past year's drinking drivers, accounting for about 46% of all drinking-and-driving trips reported in that survey. "Problem drinkers" were defined in that survey as having two or more positive responses to the CAGE instrument (Ewing, 1984), or having consumed five or more drinks on four or more days in the past month, or having consumed nine or more drinks (eight for females) on at least one occasion in the last month. These problem drinkers accounted for 343 to 491 million drinking-and-driving trips reported in 2001.

3.3. Arrests and Fatal Crashes of Alcohol-Intoxicated Drivers

In recent years, between 1.4 and 1.5 million drivers have been arrested annually for driving while intoxicated (DWI) or driving under the influence (DUI) (FBI, 2009). This is more people than are arrested each year for larceny or theft, assaults, weapons charges, or vandalism. About the same number of people is arrested each year for drug abuse violations as for DWI. In 2006, the DWI arrest rate was about one DWI arrest for every 138 licensed drivers in the United States. When combined with drinking-and-driving surveys, this amounts to one DWI arrest for every 772 reported episodes of driving after drinking, one DWI arrest for every 88 reported episodes of driving over the BAC limit, and one DWI arrest for every six stops by police for suspicion of DWI (Zador, Krawchuk & Moore, 2000).

Jones and Lacey (2000) summarized the state of knowledge on repeat DWI offenders and concluded that repeat offenders have many of the characteristics of first-time offenders. Some of these common characteristics include a high BAC at the time of arrest, alcohol dependency indications, drinking at multiple locations, and experiencing other problems related to alcohol. Kennedy et al. (1993), in an earlier review of the literature on convicted drinking drivers, reported that 80-95% of DWI offenders were males, 70-80% were aged 25-45, and 35-60% reported usually drinking 5 or more drinks at a session. In a NHTSA study, Fell (1992) showed that drivers with prior DWI convictions were overrepresented as drivers in fatal crashes by a factor of 1.8, but that only 1 out of 7 intoxicated drivers in fatal crashes had a prior DWI conviction within the past three years.

Alcohol involvement in fatal crashes (any driver with a BAC of 0.01% or greater) in 2007 was more than three times higher at night (6 p.m.–6 a.m.) than during the day (6 a.m.–6 p.m.) (62% versus 19%). Alcohol involvement was 35% during weekdays compared to 54% on weekends. Nearly one in four drivers (23%) of personal vehicles (e.g., passenger cars or light trucks) and more than one in four motorcyclists (27%) in fatal crashes were intoxicated (i.e., had a BAC equal to or greater than the 0.08% illegal limit in the United States). In contrast, only 1% of the commercial drivers of heavy trucks had BACs equal to 0.08% or higher. The 21-24 age group had the highest proportion (35%) of drivers with BACs≥0.08%, followed by the 25-34 age group (29%). The oldest and the youngest drivers had the lowest percentages of BACs ≥ 0.08%: those aged 75 or older were at 4%, and those aged 16-20 were at 17% (Fell, et al., 2009).

There were 55,681 drivers involved in fatal crashes in 2007 that resulted in 41,059 deaths. Twenty-two percent of these drivers were legally intoxicated (BAC ≥0.08%), whereas 14% had BACs exceeding 0.15% (see Figure 2). Thirteen percent of these intoxicated drivers were driving a motorcycle or an off-road vehicle in the fatal crash. Only 7.5% of these intoxicated drivers had a prior conviction for DWI within the past three years. It is believed, however, that this percentage would be much higher if the look-back period for prior offenses went beyond three years (e.g., 5 to 10 years). Over 10% of these impaired drivers were younger than 21, ages at which any drinking of alcohol and driving is illegal in every state.

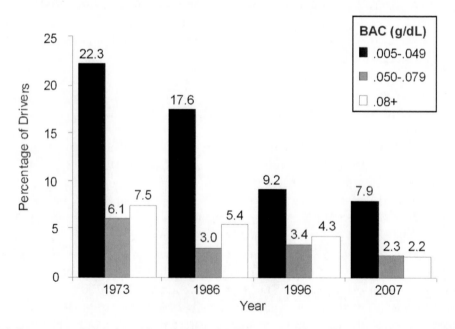

Figure 1. Percentage of nighttime drivers in three BAC categories in the four national roadside surveys.

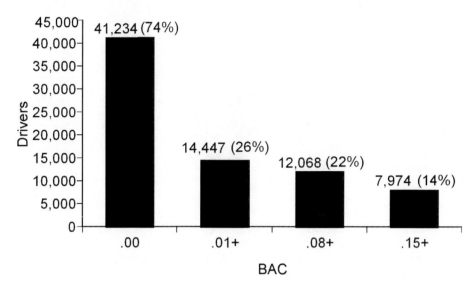

Figure 2. Drivers involved in fatal crashes by BAC level, FARS 2007 (N=55,681) (NHTSA, 2009a).

According to Voas, Romano, Tippetts & Furr-Holden (2006), one out of four drinking drivers in fatal crashes (BAC>=0.01%) are estimated to be heavy episodic drinkers, while over half are considered current normative drinkers (see Table 1).

To better understand the role of alcohol, it is useful to compare drinking and nondrinking drivers in fatal crashes, as shown in Figure 3. The first graph shows the age distribution of non-alcohol- related fatal crash involvements as a function of vehicle miles traveled (VMT) as reported by the 1990 Federal Highway Administration (FHWA) National Personal Transportation Survey. That graph takes the traditional U-shape, with underage and elderly drivers having the highest non- alcohol-related crash rates per mile driven. The common explanation for this relationship is the inexperience and risk-taking of youthful drivers and the deterioration of driving skills and, perhaps more importantly, the greater fragility of elderly drivers because of their greater susceptibility to fatal injury under certain crash conditions (Lee, 2006).

Table 1. Percent of drinking drivers in fatal crashes in five consumption categories [adapted from Voas, et al. (2006)]

Drinker Classifications	Number of Drinking Drivers in Fatal Crashes [FARS Average from 1999-2001]	Percent of Drinking Drivers in Fatal Crashes
Dependent Drinkers	1,410	11.3%
Abusive Drinkers	560	4.5%
Dependent _and_ Abusive Drinkers	270	2.2%
Heavy Episodic (Binge) Drinkers	3,170	25.3%
Current Normative (Social) Drinkers	7,110	56.8%
ALL	12,520	100%

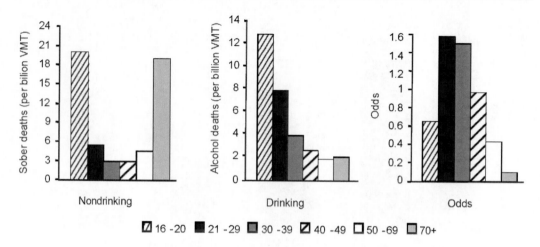

Figure 3. Drinking and nondrinking driver fatal crash rates by driver age and the odds of drinking to nondrinking driver rates, 1990-1994 [adapted from (Tippetts & Voas, 2002)].

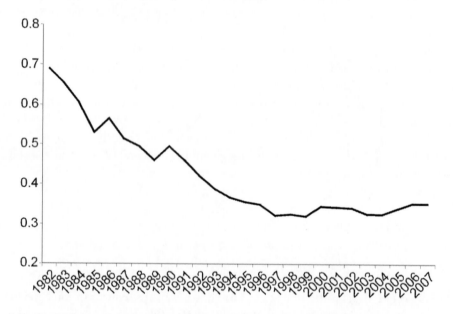

Figure 4. Ratio of drinking drivers to nondrinking drivers in fatal crashes (NHTSA, 2009a).

When, as in the second graph of Figure 3, alcohol-related rates per VMT are plotted, an L- shaped curve results with drinking-driver rates per mile driven being the highest among youthful drivers and gradually dropping with age, with the elderly least involved. The common explanation is that the driving skills of underage drivers are more vulnerable to alcohol. A somewhat different impression is provided when alcohol-involvement rates are considered as a ratio (drinking-driver rates/non-drinking-driver rates) by age group (third graph in Figure 3). This takes the shape of an inverted "U." Thus, when the involvement of underage drinking drivers in fatal crashes is related to mileage driven, their risk level is high. This is in part because their risk when sober is high. When the effect of their high risk when sober is accounted for by using that measure to normalize the data to compare across age groups, drivers 21 to 49 have a higher relative risk (odds) when drinking than do drivers under the age of 21.

The percentage of all traffic fatalities that were alcohol-related declined from 59.6% in 1982 to 41.5% in 2007, a 30% relative reduction. The percentage of all fatally injured drivers (where the testing for BAC is relatively high in the Fatality Analysis Reporting System [FARS]) who had a BAC higher than the current illegal limit in the United States (BAC $\geq 0.08\%$) dropped from 53% in 1982 to 35.5% in 2007, a 33% relative decline. A similar decline occurred for fatally injured drivers with extremely high BACs ($\geq 0.20\%$). The ratio of drinking drivers to nondrinking drivers in fatal crashes (the crash incidence ratio (CIR) reflects the effectiveness of various impaired- driving countermeasures) declined from 69% in 1982 to 35% in 2007, a 51% decline in that measure (Figure 4). This decline occurred before 1995, and the CIR has remained relatively constant for the last 15 years.

The relative risk of being involved as a driver in a fatal crash at various BAC levels (relative to a 0.00% BAC) depends on drivers' age and gender. Although not a typical case-control study, Zador, Krawchuk, and Voas (2000) selected crash data from the FARS that matched the weekend hours and the months in 1996 when a national survey of driver BACs was conducted. They showed that the relative risk of being killed as a driver in a single-vehicle crash varied considerably by age and gender, ranging from a low of 0.07 for drivers aged 35 and older at very low BACs (0.010%-0.019%) to 15,559.85 for 16-20 year old male drivers at very high BACs (0.15%+). Age and gender clearly interact with alcohol levels to influence crash risk.

These studies support several conclusions. First, the societal trends of increasing alcohol consumption, increasing instances of excessive drinking, and the stable rate of drinking and driving all suggest the need for countermeasures that go beyond the traditional enforcement and education. Second, because the minority of alcohol-intoxicated drivers in fatal crashes are problem drinkers (25%), compared to 50% who are normative drinkers. Third, because the best predictor of being a drinking driver on U.S. roads is binge drinking, countermeasures that target only repeat offenders will have limited efficacy. In combination, these characteristics of the drinking and driving problem all support a vehicle-based countermeasure based on detecting behavioral signatures of alcohol impairment.

4. SENSORS, MEASURES, AND METRICS FOR DETECTING ALCOHOL IMPAIRMENT

This study seeks to identify impairment using data from driver's own vehicles. Creating a successful vehicle-based system to detect alcohol impairment depends on the sensors and associated data. More and better data will support more precise detection, but the benefit of additional sensors must be weighed against their intrusiveness, reliability, and cost. Although EEG sensors might be able to detect changes in brain function associated with increased alcohol levels (Brookhuis & De Waard, 1993), such systems are not feasible for production vehicles. EEG sensors are expensive, produce noisy data, and require intrusive instrumentation that most drivers would not tolerate. Many other techniques do not require sophisticated sensors, such as cognitive task batteries that require people to perform a series of working memory, tracking, vigilance, and reaction time tasks (Kennedy, Turnage,

Rugotzke, & Dunlap, 1994; Kennedy, Turnage, Wilkes, & Dunlap, 1993). Unfortunately, such techniques are quite intrusive and impractical—few drivers are willing to perform these tasks every time they start a trip. Because the sensor suite must be feasible for implementation in a production vehicle, the sensors must be low-cost and non-intrusive. In addition, the sensors must be robust to a range of environmental conditions and not sensitive to the efforts of drivers to circumvent them. Recent reviews of emerging sensor technology guided our analysis of potential sensors, measures, and metrics (Pollard, Nadler, & Stearns, 2007; Ward, 2006). This section outlines the considerations for choosing sensors, measures, and metrics to detect alcohol impairment, as well as challenges associated with developing robust metrics as input to an impairment-detection algorithm. The distinction between measures (data available from sensors) and metrics (summary statistics) emphasizes the critical translation of raw data into diagnostic behavioral signatures of impairment.

4.1. Alcohol Impairment, Driving Performance, Behavior, and Safety

Figure 5 shows that alcohol has a systematic effect on driving safety, with BAC levels over 0.10% having an increasingly dramatic effect on crash risk (Blomberg, Peck, Moskowitz, Burns, & Fiorentino, 2005). This increased crash risk reflects both performance impairment and behavioral change. Performance impairment reflects diminished perceptual, motor, and cognitive capacity that accompanies increased BAC levels. Substantial evidence shows that alcohol diminishes performance, particularly in situations that require attention to be divided across multiple activities. Behavioral change reflects the increase in the willful engagement in risky behavior (e.g., speeding) with increased BAC levels. Both performance impairment and behavioral change can increase crash risk.

The degree to which performance impairment or behavioral change influences the crash risk in Figure 5 is difficult to determine, although the conditions surrounding crashes suggest that both contribute. Crash data show that alcohol-impaired drivers are more likely to be exceeding the speed limit at the time of the crash than those who are not alcohol-impaired. Similarly, crash risk for those over the age of 35 with a BAC between 0.08% and 0.10% is 11.4 times that of a sober driver, but the same BAC leads to a crash risk 51.9 times that of a sober driver for male drivers aged between 16 and 20 (Zador, Krawchuk, & Voas, 2001). The differential effect of alcohol on these two groups of drivers suggests that the influence of alcohol on crash risk depends on more than its effect on performance impairment alone.

Hundreds of studies have explored the cognitive impairments associated with alcohol. Several major reviews document these results (Ogden & Moskowitz, 2004). First, alcohol does not always lead to a statistically reliable effect on performance, but increasing BAC levels increase the likelihood of detecting an effect. Sixty percent of all studies show an effect for BAC levels over 0.09% (Evans, 1991). The difficulty in detecting the effect of alcohol at lower BAC levels suggests that people can be quite effective in adapting to low levels of alcohol impairment in low-demand situations by investing additional resources or adjusting their response strategies. Second, alcohol can affect performance at levels as low as 0.03% BAC. Situations that demand precise and timely responses show performance impairments even at low BAC levels. Third, alcohol affects the various cognitive processes differentially: perception and vigilance tasks are much less sensitive compared to selective and divided attention tasks or those associated with high stimulus-response complexity. This differential sensitivity suggests that detecting alcohol impairment would be most effective in complex and demanding driving situations that engage selective and divided attention.

Several reviews of alcohol-related performance impairment show a consistent set of findings regarding the differential effect of alcohol on task types. For studies examining impairment due to low levels of alcohol (<0.05% BAC), 70-80% show an effect of alcohol for complex tasks compared to only 33% that show an effect for simple tasks (Holloway, 1994). Simple, automatic tasks are relatively unaffected, and complex, controlled tasks are relatively sensitive to low levels of alcohol (<0.05% BAC). Complex, controlled tasks are those that require simultaneous attention to multiple task elements and lack a consistent stimulus-response mapping. Simple, automatic tasks are those that require response to a single stimulus that is paired with a single pre-defined response that the person has extensive experience selecting. Psychomotor processes associated with balance are relatively automatic and are affected only at relatively high BAC levels. In contrast, the frequency, duration, and distribution of eye fixations tend to be quite sensitive to alcohol, reflecting the sensitivity of divided attention to alcohol impairment (Moskowitz, Ziedman, & Sharma, 1976). Similarly, simple reaction time tasks are relatively unaffected and choice reaction time tasks tend to be sensitive (Tzambazis & Stough, 2000). Overall, alcohol impairs performance most strongly in situations that require people to integrate information from complex scenes, attend to multiple elements, generate responses that are contingent on multiple conditions, and coordinate multiple responses.

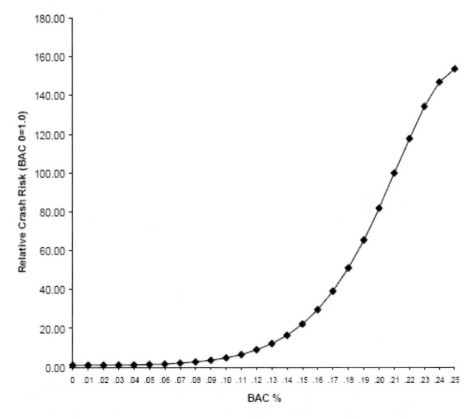

Figure 5. The effect of increasing levels of blood alcohol content (BAC) on motor vehicle crash risk (Blomberg, et al., 2005).

This simple description of alcohol impairment has important implications for selecting measures and metrics to detect alcohol impairment. First, measures depend on the driving scenarios, so scenario development is critical to gathering sensitive measures. Scenarios that involve common driving tasks occurring in isolation (e.g., lane keeping on a straight road) are likely to be less sensitive than less-practiced tasks that require distribution of attention to multiple aspects of the driving environment (e.g., adjusting to the speed of a lead vehicle on a twisting road), or coordination of vehicle control and secondary tasks (e.g., selecting a song from a CD). This dependence on driving scenarios suggests that relating measures to driving context might be an effective way to enhance their sensitivity. The second general implication is that metrics derived from multiple measures are likely to be more sensitive than metrics derived from individual measures. Alcohol impairment is most apparent in complex tasks or tasks that demand coordination on multiple dimensions. Integrating measurements from multiple aspects of driving is likely to reflect alcohol-related impairment that might not be apparent from metrics based on individual measures, a theme we will return to in defining the algorithms.

4.2. Considerations for Choosing Alcohol-Sensitive Measures and Metrics

Several critical design considerations for a vehicle-based system govern the selection of impairment metrics:

1. Sensitivity – The measure must be sensitive enough to account for a large portion of the variation in performance due to alcohol.
2. Specificity – Ideally, the measures should be indicative of alcohol impairment and not other sources of impairment, such as distraction and drowsiness, or roadway situations that might disrupt driving performance, such as work zones.
3. Availability – The measures must be accessible continuously during all driving conditions such that impairment monitoring is continuous.
4. Acceptability – The measure must appear to be specific to alcohol impairment and safety as well as not perceived as being invasive (such as speed data being used to detect impairment and enforcing speeding fines despite lack of driver impairment).
5. Practicality – The technology required for sensing the measurement data must be easily integrated into the vehicle at minimal cost and must operate robustly in all driving environments.

Of these five considerations, specificity is perhaps the most challenging to address. Specificity refers to the ability of measures to differentiate alcohol from other types of impairment. Specificity is particularly challenging because the underlying mechanisms of alcohol impairment are similar to other types of impairment. Alcohol is a central nervous system depressant, so it can induce drowsiness, which may also be manifest in people who have not had sufficient sleep. In fact, drowsiness-related driving impairment and alcohol-related driving impairment are comparable in how they affect some driving performance metrics, such as lateral control (Arnedt, Wilde, Munt, & MacLean, 2001). Furthermore, many pharmaceuticals also induce drowsiness and can impair driving performance (Lenne, Dietze, Rumbold, Redman, & Triggs, 2003; Linnoila & Mattila, 1973; Vanakoski, Mattila, &

Seppala, 2000). Because drowsiness- related impairment poses a serious threat to driving safety, differentiating alcohol and drowsiness impairment may not be necessary because both might merit similar countermeasures.

The challenge of distinguishing between drowsiness and alcohol impairment was addressed in a simulator study. Peters and van Winsum (1998) report two studies involving a highway and an overtaking scenario to examine driver impairment in relation to fatigue and alcohol. In one study, data from eight drivers were used to examine the reliability of a driver state monitor to classify drowsiness. These participants drove on two separate days after working a full night shift. On the first day, they drove for 30 minutes along a simulated highway and overtook a slower lead vehicle. Data from this initial drive were used to train a classification algorithm and were also used to define the baseline condition. On the second day, the same subjects drove a longer version of the highway route with more overtaking scenarios for a total of three hours. Data from this longer drive were used to define the fatigue condition. In a separate study, data from ten participants were used to examine the reliability of the system to classify alcohol impairment. These drivers participated in three separate drives on different days. In the initial drive, they drove for 30 minutes along a simulated highway and overtook a slower lead vehicle. Data from this initial drive were used to train the classification algorithm and also were used to define the baseline condition. For the next two drives, the same participants drove the same route twice on separate days; that is, once without alcohol and once after consuming alcohol to reach a BAC of 0.05%.

For both these studies of drowsiness and alcohol impairment, the classification algorithm was derived using data from vehicle sensors that produced measures of brake pressure, lateral position (mean, standard deviation, min), steering wheel position (standard deviation), accelerator pedal position, vehicle speed (mean, standard deviation), engine RPM, eye blink activity, and time to line crossing. In addition to an analysis of the classification performance of the system, this study also examined the similarity between the patterns of driving behavior that were derived for the normal and impaired conditions. Notably, the cases of normal and fatigued conditions overlapped in 21% to 45% of cases. The degree of overlap was even greater between the normal and alcohol conditions (41% to 65%). This suggests that (1) drowsiness has a more distinct effect than alcohol on driving, and (2) alcohol-impaired driving can be similar to unimpaired driving. Such conclusions suggest that both drowsiness and alcohol impairment can be detected, and that drowsiness might produce a similar signature to that of low levels of alcohol.

Not only do impairment from drowsiness and alcohol share a resemblance, but these two sources may combine to compound impairment. As an example, Roehrs, Beare, Zorick, & Roth (1994) examined the effect of combining alcohol and drowsiness on lateral control in a driving simulator. In this study, twelve male subjects participated in four counter-balanced test conditions: (1) full sleep with eight hours of sleep and sober; (2) full sleep and alcohol to reach BAC 0.05%; (3) partial sleep deprivation with four hours sleep and sober; and (4) partial sleep and alcohol. The results indicated that alcohol consumption increased levels of drowsiness (as measured by the Multiple Sleep Latency Test) even in the full sleep condition. Moreover, alcohol reduced lateral control, as measured by increased lane departures, when combined with drowsiness in the partial sleep deprivation. Thus, this study demonstrated the potential for a synergistic effect of alcohol in combination with drowsiness such that the risk of a crash from a combination of these factors may be higher than in the presence of either factor alone.

Another study examined the combined effect of alcohol and fatigue in car-following and lane- keeping scenarios on a two-lane roadway (Fairclough & Graham, 1999). Four matched groups of male subjects (N = 64) completed two 40-minute drives along this route: (1) control group with no sleep deprivation and a placebo drink; (2) partly sleep-deprived group with four hours sleep on night before test, placebo drink; (3) full sleep-deprived group with no hours sleep on night before test and a placebo drink; and (4) alcohol group with normal night sleep before test and an alcohol drink to yield average 0.08% BAC. This study examined performance in terms of headway and lateral control measures. The measures of lateral control were most sensitive to the impairment effects of alcohol and drowsiness. Drowsiness and alcohol both resulted in significantly more lane departures than normal driving. Moreover, drowsiness resulted in significantly less steering activity compared to both the normal and alcohol conditions. This suggests that drowsiness-related performance decrements resulted from inattention and inaction, whereas the impaired lane keeping with alcohol may result from other mechanisms such as reduced safety margins associated with the willingness to take greater risks, or ineffective steering control. Such differences may point to algorithms that differentiate impairment due to alcohol and drowsiness, but are beyond the scope of the current study.

The combined evidence of these studies supports several important conclusions. First, alcohol and drowsiness can both undermine driving performance in a way that is reflected in easily measured variables, such as steering activity and lateral position. Second, the effects of drowsiness are similar to and perhaps greater than those of alcohol, making it difficult to create an algorithm specific to alcohol and making it likely that an algorithm might correctly identify impairment associated with drowsiness even when the driver has not been dosed with alcohol. Third, the combined effect of alcohol and drowsiness may be more than either alone, which diminishes the practical importance of differentiating between drowsiness and alcohol impairment. For many interventions, the need for an algorithm that is specific to alcohol may be outweighed by the benefit of detecting impairment independent of its cause.

4.3. Measures and Metrics from Simulator and On-Road Studies

When reviewing behaviors that could be assessed to support the detection of impairment, it is important to differentiate between a behavioral *measure* and an impairment *metric*. A measure refers to data that is sensed directly, such as speed or lane position. A metric refers to the summary statistic used to characterize behavior over time or space, such as median or standard deviation. Unfortunately, there is no consistency in the literature in terms of either the choice of measure or metric to quantify driver impairment associated with alcohol (Brookhuis, De Waard, & Fairclough, 2003; Ogden & Moskowitz, 2004; Tzambazis & Stough, 2000; Zador, et al., 2001). For this reason, it is difficult to use previous research to develop a coherent picture of the effect of alcohol on driving behavior and to identify the most sensitive behavioral measures of impairment.

Considering the limits of the current research base, Table 2 provides a preliminary assessment of measures of impairment. This evaluation speculates on the compliance of each measure with each proposed criteria. As a result, it is apparent that some measures are compliant (●) while others are non-compliant (○) or have an ambiguous status (◉). For example, it is apparent that all of the measures are sensitive to impairment, although time headway (TH), stability of lateral position, and steering wheel activity have been more consistently associated with impairment effects of alcohol in the published research. However, headway measures presume the presence of a lead vehicle, which may not always be available. In contrast, speed, lateral position, and steering wheel activity are continuously available.

All of these measures would appear to have face validity in terms of intuitive relevance to impairment and safety (with perhaps the exception of steering wheel activity) and so might be understandable and acceptable to drivers. Despite this, speed may not be an acceptable measure if the public is averse to the perception that such data would be used for police enforcement of speeding fines. Finally, speed and steering wheel activity can be considered to be practical measures to the extent that effective and inexpensive sensor technologies exist. Sensors also exist to make headway and lateral position measures practical, although they tend to be more expensive and there remains some debate about the type of technology to be used for particular applications.

A range of sensors that go beyond measuring driver control inputs and vehicle state are becoming feasible for inclusion in production vehicles. These include video cameras for measuring head position and tracking eye movements. Pressure transducers in the seat can also measure movement and posture. Such sensors provide a non-intrusive, continuous, and potentially sensitive set of measures of driver impairment. For example, eyelid closure has long been recognized as a sensitive measure of drowsiness (Bergasa, Nuevo, Sotelo, Barea, & Lopez, 2006; Grace & Suski, 2001), and eye movements provide a promising indicator of distraction-related impairment (Liang, Reyes, & Lee, 2007a, 2007b) and are sensitive to alcohol (Marple-Horvat, et al., 2008; Moskowitz, et al., 1976). Eye tracking, particularly horizontal gaze nystagmus, has long been known to be sensitive to alcohol; however, eye-tracking systems that might be incorporated into production vehicles do not have the precision to estimate this metric but may support other metrics such as gaze concentration. Physiological measures are generally infeasible because they currently require a degree of intrusive instrumentation that undermines their acceptance and practicality. As vehicle automation and collision warning technology becomes more common, measures of headway and lane position will become increasingly available.

Table 2. Performance measures of driving for impairment detection

Measure	Sensitivity	Availability	Acceptance	Practicality
Steering wheel activity	●	●	◉	●
Vehicle lateral position	●	●	●	◉
Eye-tracking measures	●	◉	●	◉
Postural stability	◉	●	◉	●
Time headway (TH)	●	○	●	◉
Vehicle speed	◉	●	○	●
Time to collision (TTC)	◉	○	●	◉
Physiological measures (e.g., EEG)	●	○	○	○

Filtering the potential measures according to the four criteria for real-time detection of driver impairment identifies several measures as promising candidates. Lateral control measures, including the steering wheel, and measures of longitudinal control that are not dependent on lead vehicles (e.g., speed, pedal input) may be the most viable measures from which to compute performance metrics to quantify impairment – especially when considering the need for continuous impairment detection. Lateral control measures and metrics are more promising than longitudinal control measures.

Lateral control measures refer to the vehicle state and are partially the consequence of drivers' control input. Drivers' control input can be a more sensitive indicator of impairment because it is not filtered through the vehicle dynamics. There are three particularly promising metrics of steering control:

- Steering reversals – Mean number of steering reversals computed with a 2-degree filter (Verwey & Veltman, 1996).
- Steering error – The degree to which the steering wheel movements deviate from a smooth trajectory defined by a second-order Taylor series approximation of the steering wheel movement. Steering error provides the base data for calculating steering entropy.
- Steering entropy – The predictability of the driver's steering responses, as defined in Nakayama et al. (1999); also see Boer (2001).

Whereas these measures can be sensitive to impairment, steering entropy can be problematic given that low entropy (and few reversals) can be indicative of attention to the driving task, very effective control input, or a lack of attention with insufficient control activity. In both these cases, the amount of input is minimal or absent, which gives the impression of equivalent states, although the case of diminished input associated with lack of attention is clearly an indication of impairment.

Steering entropy (SE) has been extensively developed and applied to detecting distraction-related impairment (Boer, 2001; Nakayama, et al., 1999). Other metrics of steering behavior have also been used (e.g., reversals, standard deviation of steering position, power spectral analysis). However, most of these metrics have been used primarily to characterize impairment effects of drowsiness or distraction. Naturally, since the sedative effect of alcohol does resemble some of the characteristics of drowsy driving, these alternative measures of steering activity may also be useful. Recent research suggests SE is very sensitive to distraction, but may not be any more sensitive to alcohol than simpler metrics of steering reversals (Rakauskas, et al., 2008). However, SE may be less sensitive to contextual variables (e.g., road curvature) than some of the alternative metrics. For this reason, SE may be a promising alternative to steering reversals.

4.4. Measures and Metrics from Naturalistic Roadway Observations

Substantial research has shown that behavioral cues available to police officers monitoring drivers can be indicative of alcohol-related driving impairment (Harris, 1980). These cues include those that can be observed while the car is in motion, such as weaving, and cues that are observed after a police officer has stopped a motorist, such as difficulty exiting the vehicle. A study confirmed the validity of these cues and found that the same cues used to identify drivers at 0.10% BAC were also effective in identifying those at 0.08% BAC (Stuster, 1997). Interestingly, none of these cues proved to be indicative of BAC levels below 0.08%, despite an attempt to identify such cues. Candidate cues sensitive to low BAC levels were identified through a review of relevant literature and through interviews with police officers. Table 3 shows the cues most sensitive to alcohol impairment; those that were not sensitive to alcohol impairment, such as speeding more than 10 miles over the speed limit, have not been included.

Similar to the conclusions based on simulator and on-road experiments, Table 3 suggests that lateral control may be sensitive to alcohol and that metrics such as the standard deviation of lane position might be most effective. An automated system to detect alcohol impairment was developed and found lateral control, rather than longitudinal control, to be sensitive. Data were collected over approximately 10 seconds as drivers decelerated in a 300 ft approach lane to a checkpoint (Stuster, 1999). The difference between the maximum and minimum lateral displacement and the minimum lateral displacement were both sensitive to alcohol and, when combined, they identified 67% of those with BAC levels above 0.04% in a population of drivers in which 80% had BAC levels of 0.04% to 0.12%.

4.5. Conclusion

The measurement of driver impairment often involves the formulation of metrics based on safety-relevant measures. Typically, these measures quantify behaviors associated with operating vehicle controls and the resulting vehicle state with respect to lateral and longitudinal control. The most common behavioral measures that have a demonstrated sensitivity to impairment are metrics of steering input and lateral control. Such measures are relevant, given that increased lateral variability is related to the probability of a lane departure that is itself a precursor to numerous types of crashes, including road departures. The standard deviation of lane position and steering reversals are two particularly promising metrics. These measures are safety-relevant in that they have direct implications for vehicle control. Other indicators of impairment, such as the association of eye movements and steering wheel position, may not have such a direct relationship with vehicle control, but may be diagnostic. Safety-relevant metrics may be advantageous if the algorithm produces an output that the driver must interpret: safety-relevant metrics are more likely to be understood and accepted.

The nature of alcohol impairment suggests that metrics based on integrated measures, rather than metrics based on single measures, represent a productive and underexplored approach. Alcohol impairment is most apparent in complex, multi-task situations. Consequently, metrics that reflect the joint performance and coordination of multiple driving tasks seem most likely to be sensitive to alcohol impairment. The relationship between horizontal eye position and steering wheel angle is one such example of coordinated tasks.

Overall, there is no critical mass of evidence to recommend specific metrics. Moreover, few studies have deliberately compared various measures and metrics in terms of sensitivity to alcohol in the driving context. Based on the four criteria listed above, the most promising measures and associated metrics are:

Table 3. Cues for identifying alcohol-impaired drivers from field studies (Stuster, 1997)

Element of Driving	Behavioral Indicator (probability of impairment given observed behavior)
Lane Position Maintenance	Weaving within a lane (0.52)
	Weaving across lane lines (0.54)
	Straddling a lane line (0.61)
	Swerving (0.78)
	Turning with a wide radius (0.68)
	Drifting during a curve (0.51)
	Almost striking a vehicle or other object (0.79)
Speed Control and Braking	Stopping problems (too far, too short, or too jerky) (0.69)
	Accelerating or decelerating for no apparent reason (0.70)
	Varying speed (0.49)
	Slow speed (10+ mph under limit) (0.48)
Vigilance	Driving in opposing lanes or wrong way on one-way (0.54)
	Slow response to traffic signals (0.65)
	Slow or failure to respond to officer's signals (0.65)
	Stopping in lane for no apparent reason (0.55)
	Driving without headlights at night (0.14)
	Failure to signal or signal inconsistent with action (0.18)
Judgment	Following too closely (0.37)
	Improper or unsafe lane change (0.35)
	Improper turn (too jerky, sharp, etc.) (0.50)
	Driving on other than the designated roadway (0.80)
	Stopping inappropriately in response to officer (0.69)
	Inappropriate or unusual behavior (throwing, arguing, etc.) (0.48)
	Appearing to be impaired (0.90)
	Illegal turn (0.19)

- Steering wheel activity – Reversals and steering error
- Vehicle lateral position – Standard deviation of lane position and range
- Eye movement – Gaze concentration

5. DETECTION CRITERIA AND CONSIDERATIONS

The criterion that defines a particular level of impairment represents an important design consideration for any impairment-detection system. Indeed, assessing impairment depends on specifying a classification rule rather than simply selecting a metric or behavioral signature. There are several frameworks for establishing the thresholds that define impairment: absolute criteria, relative criteria, and pattern-based criteria. Any practical algorithm is likely to contain elements of a pattern-based algorithm because alcohol impairment is most likely to be revealed in a combination of performance decrements rather than declines on one or two measures.

5.1. Absolute Criteria

Absolute criteria classify behavior as impaired based on a threshold that is independent of the distribution of unimpaired behavior in a baseline condition. Absolute criteria reflect safety boundaries and human performance capabilities. Absolute criteria can also be set in terms of thresholds that are accepted by convention as having face validity for safety, such as excessive speed; for example, a speed threshold can be set at the speed limit plus a margin of 10%.

Based on these considerations, Brookhuis, De Waard, & Fairclough (2003) have proposed the "tentative" list of absolute criteria shown in Table 4. These criteria were derived from a subset of possible behavioral metrics from impairment conditions of alcohol, distraction, drowsiness, and visual occlusion[1]. Based on these criteria, a driver would be classified as impaired if the standard deviation of lane position changed by more than 4 cm (relative criterion) or was more than 25 cm (absolute criterion). Such diagnoses of impairment are considered valid because these criteria have been calibrated with the effects of alcohol at BAC 0.08%.

5.2. Relative Criteria

Relative criteria classify behavior as impaired based on a threshold that is defined relative to the distribution of normal behavior in an unimpaired baseline condition. Relative criteria implicitly define safety in terms of the *change* in crash risk from the baseline condition. Such criteria are determined by testing the statistical significance of the deviation between observed behaviors and the distribution of behaviors in the baseline condition. A statistically significant deviation signifies that the observed behavior is probably not representative of the baseline condition and is, therefore, presumed to represent an impaired condition. Based on these considerations, Brookhuis, De Waard, & Fairclough (2003) have also proposed the list of relative criteria shown in Table 4. The criteria listed in Table 4 represent absolute and relative decrements in driving performance based on observed driving behavior, and the criteria listed in Table 5 are based on observations of drivers in the visual occlusion paradigm. The absolute criteria in Table 5 are absolute in the sense they do not depend on the driver, but are adjusted according to the speed of the driver.

5.3. Pattern-Based Criteria

Pattern recognition criteria are based on neural nets or similar statistical approaches to generate multivariate rules that are "trained" to differentiate between normal and "impaired" states. For example, De Waard, Brookhuis, & Hernández-Gress (2001) conducted a driving simulator study with 20 drivers to evaluate the performance of an impairment classification system to detect driver distraction based on input parameters of pedal and steering control, vehicle speed, lateral position, and headway to lead vehicle.

**Table 4. Definition criteria for following too closely, straddling lanes,
and driving too fast (Brookhuis, et al., 2003)**

Measure	Absolute Change	Relative Change
Following too close:		
Time headway to lead vehicle (TTC)	<0.7 s	-0.3 s
Lane keeping:		
Steering standard deviation (SD)	>1.5	+0.5
Lateral deviation (SD) of the vehicle	>0.25 m	+0.04 m
Minimum time-to-line crossing (TLC) right	<1.3 s	-0.3 s
Minimum time-to-line crossing (TLC) left	<1.7 s	-0.2 s
Median TLC (right lane)	<3.1 s	-0.7 s
Median TLC (left lane)	<4.0 s	-1.4 s
Driving too fast:		
Vehicle speed[2]	limit +10%	+/- 20%

Table 5. Lane-keeping criteria based on visual occlusion (Brookhuis et al., 2003)

Measure	Speed (km/h)	Absolute Criteria
Standard deviation of lane position	>50	0.25 m
Standard deviation of steering wheel	At 60	1.7°
	>80-120	1.5°
Median TLC	60	6.0 s
	80	5.7 s
	100	5.0 s
	120	4.2 s
15% TLC	60	3.8 s
	80	3.5 s
	100	3.1 s
	120	2.9 s
Minimum TLC at different speeds		1.1 s

The classification system had three components, as shown in Figure 6. First, the original set of measures from the sensor data is reduced to a number of composite variables or metrics that transform and aggregate the individual measures. Second, pattern discovery is applied to identify signatures of impairment. These signatures are combinations of metrics that differentiate normal and non-normal conditions. Third, a classification algorithm (e.g., Support Vector Machine or decision tree) recognizes patterns of measured behavior and establishes thresholds that differentiate between normal and non-normal conditions. This requires a training process in which the algorithm is told the actual condition for the assigned data so that patterns and thresholds can be learned. The algorithm is then used to recognize those learned patterns indicative of normal and impaired conditions for new data sets. Note that this system in effect uses relative criteria, but unlike other conventions that apply thresholds to individual measures (see Table 4), it adopts multidimensional thresholds for derived components that characterize patterns of behavior indicative of impairment. The result of this process is the detection of non- normal conditions, such as instances of impaired driving.

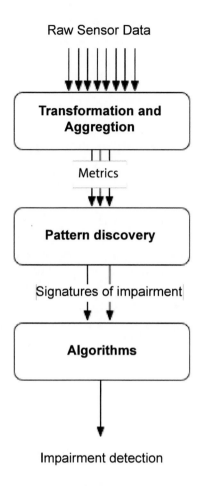

Figure 6. Representation of processes in generic classification system to classify normal and non-normal conditions.

5.4. Selecting Appropriate Criteria: Matching Criteria to Mitigations

Each of these general criteria has different strengths and weaknesses. Most generally, the absolute criteria are most easily understood and explained. In contrast, the relative and pattern- based criteria are likely to be much more effective in distinguishing impaired from unimpaired drivers because they make more complete use of the data. The relative importance of understandability and effectiveness depends on the particular mitigation that the algorithm will ultimately guide. If the algorithm drives immediate feedback to the driver, then driver acceptance of the information is critical and understandability would be relatively important. Drivers might not understand, and therefore ignore, an impairment warning based on how quickly their eyelids close, but would be more likely to understand a warning based on weaving or swerving. If the algorithm guides collision warning adaptation, then the effectiveness would be most critical. In addition, some algorithms, such as those based on driver state (e.g., postural stability and eye movements), do not have any well-defined absolute criteria and their output does not have benchmarks that can be easily understood by the driver. The primary purpose of this study is to develop an algorithm that can identify alcohol-impaired drivers, so the emphasis will be on pattern-based criteria in which multiple metrics are combined and evaluated to maximize correct detection and minimize false positives.

6. DATA COLLECTION METHODS

Algorithm development and evaluation requires detailed data from alcohol-impaired drivers in a controlled, yet realistic situation. Data were collected from drivers at three BAC levels experiencing representative driving scenarios in a high-fidelity driving simulator. The data collection involved 108 drivers from three age groups (21-34, 38-51, and 55-68 years of age) driving through representative situations on three types of roadways (urban, freeway, and rural) at three levels of alcohol concentration (0.00%, 0.05%, and 0.10% BAC). BAC dosing was limited to 0.10% due to practical and ethical reasons. The following sections summarize the data collection methods: participant population, simulator and sensor suite, experimental design, and dependent variables. Details are provided in appendices referenced in each section.

6.1. Participants

One hundred and eight participants completed all three drives of the study. Participants were healthy men and women aged 21 and older, with a valid driver's license, and who were moderate to heavy drinkers. All drivers had been licensed for at least two years and drove a minimum of 10,000 miles per year. Efforts were made to recruit a racially and ethnically diverse participant population. Participants were paid $250 for completing all study sessions. Pro-rated compensation was provided for participants who did not complete the study. Inclusion and exclusion criteria include:

- Possess a valid US driver's license
- Licensed driver for two or more years
- Drive at least 10,000 miles per year
- Restrictions on driver's license limited to vision
- Not currently taking illegal drugs or drugs that interact with alcohol
  All participants provided a urine specimen on each of the three experimental days and were screened for 10 types of drugs: methamphetamine, morphine, cocaine, marijuana, PCP, benzodiazepines, barbiturates, methadone, trycyclic antidepressants, and amphetamine. Those participants who tested positive were discontinued from participation in the study. The urine specimens of females were screened for hCG, a pregnancy hormone. Due to health concerns, those participants who tested positive for pregnancy or illicit drug use were discontinued from the study.

- Does not use any special equipment to drive, such as pedal extensions, hand brake or throttle, spinner wheel knobs, or other non-standard equipment that would limit interpretation of accelerator pedal, brake pedal, or steering inputs.
- Be a moderate to heavy drinker as determined by the Quantity-Frequency-Variability (QFV) scale. Appendix A contains the survey and scoring that defines the specific inclusion and exclusion criteria for this criterion.

Participants were recruited to fit specific age and gender characteristics and, as shown in Table 6.

Table 7 summarizes the population of participants showing the intended balance of age and gender. The distribution of drinking patterns based upon the QFV reflects that participants in this study were more likely to be heavy drinkers rather than moderate drinkers. The sleepiness scale reveals that, with the exception of the older females, drivers were more drowsy at the completion of their drive that they were before it started. Overall simulator sickness scores reflect the relative lack of adverse affects associated with driving in a simulated environment.

Table 6. Number of participants reporting to visits[3]

Group	Enrolled Visit 1 (Screening)	Passed Screening	Visit 2	Visit 3	Visit 4	Completed
Young Male	35	24	21	18	18	18
Young Female	24	20	20	18	18	18
Middle Male	27	20	20	18	18	18
Middle Female	27	20	19	18	18	18
Older Male	31	26	22	20	19	18
Older Female	21	20	19	18	18	18
Total	165	130	121	110	109	108

Table 7. Participant characteristics

Variable		Age 21-34 Male	Age 21-34 Female	Age 38-51 Male	Age 38-51 Female	Age 55-68 Male	Age 55-68 Female
Number completed		18	18	18	18	18	18
Mean age (years)		26.56	26.83	43.22	44.72	59.56	61.06
Mean height (inches)		70.65	65.53	70.61	65.35	70.14	64.76
Mean weight (pounds)		199.81	159.56	220.61	175.29	211.86	172.86
Mean body mass index		27.9	26.1	31.1	28.6	30.2	29.0
Distribution of Drinking Patterns	Moderate	2	7	4	9	6	7
	Heavy	16	11	14	9	12	11
Sleepiness scale	Pre-Drive	2.3	2.6	2.3	2.2	2.0	2.0
	Post-Drive	2.7	3.1	2.8	2.4	2.3	2.0
Simulator sickness score		8.9	18.3	19.2	21.9	17.9	15.1

Figure 7. Representation of NADS-1 driving simulator (left) with a driving scene from inside the dome (right).

6.2. Simulator and Sensor Suite

The National Advanced Driving Simulator (NADS) is located at The University of Iowa's Oakdale Campus. It consists of a 24-foot dome in which an entire car is mounted. All participants drove the same vehicle—a 1996 Malibu sedan. The motion system, on which the dome is mounted, provides 400 square meters of horizontal and longitudinal travel and ±330 degrees of rotation. The driver feels acceleration, braking, and steering cues as if he or she were actually driving a real vehicle. Each of the three front projectors has a resolution of 1600 x 1200; the five rear projectors have a resolution of 1024 x 768. The edge blending between projectors is five degrees horizontal. The NADS produces a complete record of vehicle state (e.g., lane position) and driver inputs (e.g., steering wheel position), sampled at 240 Hz.

The cab is equipped with a Face Lab™ 4.0 (Seeing Machines, Canberra, Australia) eye-tracking system that is mounted on the dash in front of the driver's seat above the steering wheel. The worst-case head-pose accuracy is estimated to be about 5°. In the best case, where the head is motionless and both eyes are visible, a fixated gaze may be measured with an RMS error of 2°.

Softflex™ (Vista Medical, Winnipeg, Manitoba) force sensor array (FSA) mats were positioned on the bottom and back of the driver's seat. Each of these mats consists of 256 piezo resistive sensors distributed in a 16 x 16 array over an area of approximately 17" x 17." These mats sample data from the sensor array a rate of 35 Hz.

An Alco-Sensor IV (Intoximeters, Inc., St. Louis, MO) breath alcohol-testing instrument was used to measure participants' BAC. The hand-held sensor uses a fuel cell to determine BAC level. The system is approved by the US DOT for evidential use and exceeds the federal model specification for traffic enforcement and Omnibus Breath Alcohol Testing. The system is designed to measure BAC levels from 0.00% to 0.40% with drift of less than 0.005% BAC over several months. The system was checked at least every other day for calibration and recalibrated using an approved dry gas standard.

6.3. Driving Scenarios

Each drive was composed of three nighttime driving segments. The drives started with an urban segment composed of a two-lane roadway through a city with posted speed limits of 25 to 45 mph with signal-controlled and uncontrolled intersections. An interstate segment followed that consisted of a four-lane divided expressway with a posted speed limit of 70 mph. Following a period in which drivers followed the vehicle ahead, they encountered infrequent lane changes associated with the need to pass several slower-moving trucks. The drives concluded with a rural segment composed of a two-lane undivided road with curves. A portion of the rural segment was gravel. These three segments mimicked a drive home from an urban bar to a rural home via an interstate. Events in each of the three segments combined to provide a representative trip home in which drivers encountered situations that might be encountered in a real drive. Scenario events are summarized in Table 8.

Throughout the urban section, a series of potential hazards required drivers to scan the roadside. These hazards included pedestrians, motorbikes, and cars entering and exiting the roadway. These hazards had paths that would cross the driver's path if they were to remain on their initial headings. There was an instance where a pedestrian crossed the driver's path well in front of the driver.

Because each participant drove three times, once for each BAC level, three scenarios with varied event orders were required to minimize the learning effects from one drive to the next. For each of the three scenarios, there were the same number of curves and turns, but the order of the curves varied. For example, the position of the left turn in the urban section varied so that it was located at a different position for each drive. Additionally, the order of the left and right rural curves varied between drives. The scenario specification in Appendix B provides additional details concerning the differences between the three drives and the events.

6.4. Experimental Design and Independent Variables

A 3 x 2 x 3 between-between-within subjects design exposed six groups of participants to three BAC levels. Between-subject independent variables were age group (21-34, 38-51, and 55-68 years) and gender. The within-subject independent variable was BAC (0.00%, 0.05%, and 0.10%).

Three factors motivated the choice of the age ranges. The first factor was that only those who could legally drink in the state of Iowa would be included. Therefore, enrollment in the study was restricted to those 21 years of age or older. The second factor was that to the extent possible, the entire spectrum of adults who drink and drive should be included, which motivated including a group with maximum age of approximately 70. The third factor was that the age ranges should be uniform, with equal spacing between them. Based on these requirements, the age groups were 21-34, 38-51, and 55-68, so that each group had a range of fourteen years.

6.5. Procedure

Participants were recruited with newspaper ads, internet postings, and referrals (see Appendix C for recruitment materials). An initial telephone interview determined eligibility for the study. Applicants were screened in terms of health history, current health status, and use of alcohol and other drugs (see Appendix D). The Cahalan, Cisin, and

Crossley Quantity-Frequency-Variability (QFV) scale was used to determine whether applicants were moderate drinkers or heavy drinkers and eligible for participation in the study (see Appendix A). The Audit survey was used to exclude chronic alcohol abusers (see Appendix E). Pregnancy, disease, or evidence of substance abuse resulted in exclusion from the study. Participants taking prescription medications that interact with alcohol were also excluded from the study.

Table 8. Summary of events for each of the three segments of the drive

Segment	Event Name (number)	Description	Approximate Duration (seconds)
Urban	Pull Out (101)	Pull out of parallel parking spot into traffic	30
	Urban Drive (102)	Drive on a narrow two-lane road with traffic and parked vehicles	45
	Green Light (103)	Navigate green traffic light on urban two-lane road with parked vehicles along the road, oncoming traffic, traffic behind driver	30
	Yellow Dilemma (104)	Navigate yellow light dilemma on urban two-lane with parked vehicle, oncoming traffic, traffic behind driver	75
	Left Turn (105)	Left turn at signalized intersection (no green arrow, no dedicated turn lane), oncoming traffic, variety of gaps	80
	Urban Curves (106)	Three curve segments of mixed radius of curvature	180
Freeway	Turn On Ramp (201)	Turn right onto interstate on-ramp	30
	Merge On (202)	Merge onto interstate	50
	Following (203)	Intermittent slower-moving truck traffic in the driving lane and a single slow moving passenger vehicle in the passing lane, interleaved with a CD- change distraction task	180
	Merging Traffic (204)	Approach second interchange, interact with traffic merging into interstate	60
	Interstate Curves (205)	Navigate three curves on interstate	185
	Exit Ramp (206)	Take exit ramp off interstate	30
Rural	Turn Off Ramp (301)	Turn right from ramp onto rural two-lane road	30
	Lighted Rural (302)	Lighted two-lane rural road, 55 mph	90
	Transition to Dark Rural (303)	Straight roadway that transitions between lighted and unlit	20

Segment	Event Name (number)	Description	Approximate Duration (seconds)
	Dark Rural (304)	Unlit straight and curved road, segments, center and road edge marking are faded and the road surface is grayish. Data from the Hairpin Curve is not included in this.	270
	Dark Rural Hairpin Curve (304)	A hairpin turn and a vertical curve located within the Dark Rural (304). Data for Dark Rural and Hairpin Curve are mutually exclusive.	30
	Gravel Transition (305)	Transition to gravel surface on straight road	30
	Gravel Rural (306)	Gravel road (straight and curves)	90
	Driveway (307)	Pull into driveway with gravel	30

Each participant participated in four sessions, the last three separated by one week. Order of target BAC levels and scenario event sequence were counterbalanced across participants, as shown in Appendix F. The time of day of each of the three sessions was the same for a given participant.

Appendix F includes the full experimental protocol. On study Visit 1 (screening), upon arrival at the NADS, each participant gave informed consent to participate in the study and received a copy of the signed informed consent form (see Appendix I). They then provided a urine sample for the drug screen and, for females, the pregnancy screen. During a five-minute period following these activities, the participant sat alone in the room where subsequent measurements of blood pressure, heart rate, height, and weight were made. Cardiovascular measures within acceptable ranges (systolic blood pressure = 120 ± 30 mmHg, diastolic blood pressure = 80 ± 20 mmHg, heart rate = 70 ± 20) confirmed eligibility for the study. If participants met study criteria, they were then administered a breath alcohol test and verbally administered the QFV (see Appendix J) and the Audit Survey (see Appendix E) to further confirm eligibility. If participants met study criteria, they completed demographic surveys. These surveys included questions related to crashes, moving violations, driver behavior, drinking, and driving history (see Appendix K). Participants viewed an orientation and training presentation (see Appendix L) that provided an overview of the simulator cab and the in-cab task they were asked to complete while driving. Participants then completed the practice drive (see Appendix G for in-cab protocol and Appendix H for control room logs) and completed surveys after their drive about and how they felt and about the realism of the simulator (see Appendices M and N). The practice drive included making a left hand turn, driving on two- and four-lane roads, and practicing the CD changing task. Date and time for Visits 2, 3, and 4 were confirmed with participants at the completion of this visit.

During Visits 2, 3, and 4 all participants completed a urine drug screen and, for females, a pregnancy screen to confirm eligibility for the study. Participants waited for five minutes following these activities during which the participant sat alone in the room where subsequent measurements of blood pressure and heart rate were obtained to determine study eligibility. If participants met study criteria, they then received a breath alcohol test, the QFV, and the

Audit Survey to further confirm eligibility. If eligible to continue, the time and duration of last sleep, and time and contents of last meal were recorded (Appendix O). Age, gender, height, weight, and drinking practice were used to calculate the alcohol dose. The Sahlgrenska Formula was used to estimate body water for each participant in order to calculate the amount of alcohol and juice required to reach the target BAC (Equation 1 and Equation 2). Participants were served three equal-sized drinks at 10-minute intervals and were instructed to pace each drink evenly over the 10-minute period. NADS staff monitored the participants periodically throughout the drinking period to ensure an even pace of drinking.

Equation 1. Sahlgrenska formula for body water.

$$Body\ Water\ (liters) = \begin{cases} if\ Male, -10.759 + 0.192 \times Height(cm) + 0.312 \times Weight(Kg) - 0.078 \times Age(years) \\ if\ Female, -29.994 + 0.294 \times Height(cm) + 0.214 \times Weight(Kg) - 0.0004 \times Age(years) \end{cases}$$

Equation 2. Vodka volume formula.

$$Amount\ of\ Vodka\ (mL) = \frac{DesiredPeakBAC + \frac{\frac{Total\ Absorption\ Time}{60 \times Standard\ Clearance\ Rate}}{H_2O\ in\ Blood} \times \frac{Sahlgrenska\ Body\ Water \times 10}{Specific\ Gravity\ of\ Alcohol}}{0.4}$$

$$\begin{bmatrix} Specific\ Gravity\ of\ Alcohol\ = 0.79 \\ Sahlgrenska\ Body\ Water\ = Equation\ 1 \\ H_2O\ in\ Blood\ = 85\% \\ Desired\ Peak\ BAC\ = 0.115\%\ or\ 0.065\% \\ Standard\ Clearance\ Rate\ = 0.017 \end{bmatrix}$$

On the days when participants were dosed to achieve 0.10% and 0.05% BAC, the amount of alcohol consumed was calculated to produce a peak BAC of 0.115% or a peak BAC of 0.065%. On the 0.00% peak BAC day, the drink consisted of one part water and 1.5 parts orange juice. Each of the glasses had its rim swabbed with vodka and 10 ml of vodka was floated to produce an initial taste and odor of alcohol.

Sixteen minutes after the end of the third drink, BAC measurements were taken at two- to five- minute intervals until the target BAC (± 0.005%) was reached. BAC values were plotted to identify the peak BAC. Each data point includes the measurement error of 0.005% associated with the accuracy of the sensor. A sample plot for identifying when BAC has peaked and begun to decline to the point where the participants could go into the simulator with a BAC of 0.10% is provided in Appendix P. Prior to insertion in the simulator, a participant must have had at least two declining data points and been within the target range, or below if the target was not reached. Peak BAC was expected 30 minutes after the end of the third drink.

When the target BAC was reached, the participants drove in the NADS. All data were collected as the BAC declined to minimize extraneous variation associated with the effect of rising and falling BAC levels and to represent the most likely situation under which alcohol-impaired driving occurs. As soon as the simulator returned to the dock and the participant exited the simulator (within 5 minutes of completing the drive), a BAC measurement was obtained, followed by an SFST (see Appendix Q). The individuals conducting the SFST were trained according to NHTSA's guidelines. The primary individual leading the training and administration of the SFST was a former police officer with 20 years of experience in law enforcement. The Stanford Sleepiness scale was also administered before and after each drive (see Appendix R).

Participants were not informed of their measured BACs until their participation in the study was completed. On all experimental days, the participants were transported home after their BAC dropped below 0.03%. Measurements of BAC were taken every 20 minutes after exiting the simulator to indicate when the participant was to be transported home. In the case of the 0.00% BAC condition, participants were held for at least three measurements before being transported home. At the end of Visit 4, participants were debriefed (see Appendixes S and T) and paid $250. Pro-rated compensation was provided for participants who did not complete the study.

6.6. Dependent Variables

Each drive consisted of 19 distinct events, so the specific dependent variables were not constant across the drive. The primary measures examined across the drive were standard deviation of lane position, average speed, and standard deviation of speed. Average speed over a segment was calculated as the mean velocity irrespective of the speed limit as was the standard deviation of speed. The scenario specification describes the dependent variables for each event (see Appendix B).

A critical step in developing algorithms concerns the translation of measures into alcohol-sensitive metrics. The data collection produced an archive of more than 100 hours of driving data, sampled at 240 Hz, across 19 distinct events. Metrics of driving performance are not invariant over roadway situations. The road type, traffic situations, and the particular maneuver the driver happens to be performing all make it important to consider the metrics in the context of the roadway situation. Alcohol-related metrics may differ on a segment-by-segment basis and even on a minute-by-minute basis. Lane position averaged over a trip might mask lane-straddling behavior on a freeway that did not occur on an urban arterial during the same trip. A challenge in aggregating the raw data is to avoid combining qualitatively different data for different behaviors and situations, for this reason data were aggregated over each event.

The following sections describe the algorithm development and evaluation that builds on the data reduction. The primary objectives associated with the algorithm development and analysis include:

- Understand how driving-related metrics reflect the impairment associated with BAC at 0.05% and 0.10%
- Determine the robustness of these metrics with respect to individual differences such as age, and gender, and roadway situation
- Develop algorithms to detect alcohol-related impairment
- Compare robustness and timeliness of metrics and algorithms

These sections serve two general purposes: to describe the effect of alcohol on driving and to assess how well algorithms can identify BAC levels over the legal limit.

7. ALCOHOL LEVELS AND DRIVING PERFORMANCE

This section describes participants' BAC levels and the effect of these levels on three standard driving performance measures: standard deviation of lane position, mean speed, and speed deviation. This section first describes the stability of driving performance across the three drives, indicating where drivers might have adapted to the scenario with repeated exposure to the events. Although the independent variables were counterbalanced to minimize learning effects, substantial changes in response to events over time might undermine the sensitivity and limit the generalization of the results. This section then describes how the three dependent measures (lane deviation, average speed, and speed deviation) varied as a function of BAC, age, and gender. This analysis also describes the sensitivity of driving performance during each event to the BAC levels. Overall, this section addresses the following objectives:

- Understand how driving-related metrics reflect the impairment associated with BAC at 0.05% and 0.10%
- Determine the robustness of these metrics with respect to individual differences such as age, and gender, and roadway situation

Based on previous research concerning the effect of alcohol on driving performance, the analysis focuses on three measures, which are summarized in Table 9. It is not meaningful to calculate these measures for some events, such as pulling out of a parking spot (Event 101), which is indicated in the table.

7.1. Blood Alcohol Concentrations of Participants

Table 10 shows the BACs that were obtained pre-drive, post-drive, and the pre-post drive average. The values for the conditions where drivers were not dosed were 0.00% BAC. The dosing and testing procedures effectively produced the intended experimental conditions. The median BAC for the 0.05% condition was 0.048%, and the median BAC for the 0.10% condition was 0.095%. Figure 8 shows the individual data points that underlie the data in Table 10. Points under the diagonal line represent cases where the post-drive BAC was lower than the pre-drive BAC. These points represent cases where the drivers had declining BAC during the drive, which was a goal of the dosing protocol that was achieved in 204 of the 216 cases.

7.2. Driver Adaptation to Events with Repeated Exposure

Ideally, driver response to each event would be unaffected by previous exposure to the events— their response would be similar across the three drives. The following analysis examines the extent to which the results departed from that ideal.

Table 9. Events for which Standard Deviation of Lane Position, Average Speed, and Speed Deviation were analyzed

Event Name (number)	Lane Deviation	Average Speed	Speed Deviation
Pull Out (101)			
Urban Drive (102)	√	√	√
Green Light (103)	√	√	√
Yellow Dilemma (104)	√	√	√
Left Turn (105)		√	
Urban Curves (106)	√	√	√
Turn On Ramp (201)			
Merge On (202)			
Following (203)			
Merging Traffic (204)			
Interstate Curves (205)	√	√	√
Exit Ramp (206)		√	
Turn Off Ramp (301)			
Lighted Rural (302)	√	√	√
Transition to DarkRural (303)	√	√	√
Dark Rural (304)	√	√	√
Dark Rural hair pin curve (304)	√	√	√
Gravel Transition (305)		√	√
Gravel Rural (306)	√	√	√
Driveway (307)			

Table 10. Summary of BAC levels for the two experimental conditions

Test Time	0.05% BAC (N = 108)			0.10% BAC (N = 108)		
	M	SD	Median	M	SD	Median
Pre-drive	0.053	0.005	0.054	0.098	0.009	0.102
Post-drive	0.042	0.006	0.043	0.088	0.009	0.090
Mean	0.047	0.005	0.048	0.093	0.008	0.095

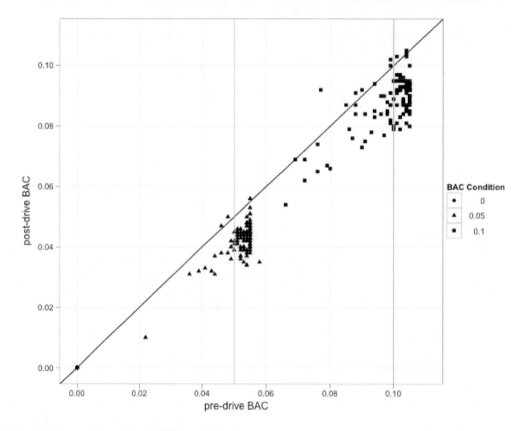

Figure 8. Pre- and post-drive BAC levels for the two experimental conditions.

Figure 9 shows lane deviation, average speed, and speed deviation by drive number across scenario events. No lane deviation measures had statistically reliable differences as a function of drive number (1, 2, or 3). The variation between events is much greater than the variation across drives. The standard deviation of lane position is stable over drives.

For average speed, there were statistically significant differences as a function of drive number in Interstate Curves (205), Exit Ramp (206), Transition to Dark Rural (303), Gravel Transition (305) and Gravel Rural (306) events. Although statistically significant, the differences between drives are small relative to the differences between events and only in Gravel Rural (306) is the difference substantial. This event consisted of driving on a gravel road, which might not have been a routine driving experience for some participants, and the increase may be explained by increased comfort with the task.

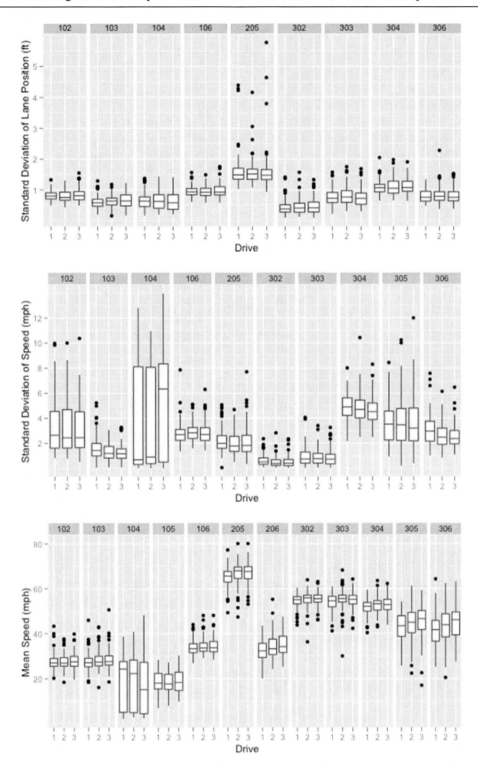

Figure 9. Standard deviation of lane position, average speed, and speed deviation by drive number across scenario events for Urban Drive (102), Green Light (103), Yellow Dilemma (104), Urban Curves (106), Interstate Curves (205), Lighted Rural (302), Transition to Dark Rural (303), Dark Rural (304), Gravel Transition (305), and Gravel Rural (306).

For the speed deviation measures, there were statistically significant differences as a function of drive number in the Green Light (103) and Gravel Transition (306) events. Although not statistically significant, Yellow Dilemma (104) had the largest change in speed variation. This difference likely stems from more participants noticing and braking for the yellow light on subsequent drives.

Overall, these analyses show generally stable driving behavior with a minimal amount of learning or adaptation over the three drives. Several events, such as the gravel road and the several urban events show moderate changes in speed selection and modulation. However, these changes are small relative to the differences between events.

7.3. Effect of BAC Levels on Driving Performance across Roadway Situations

The purpose of this analysis is to assess the overall sensitivity of common driving metrics to alcohol and to assess how robust they are to differences between roadway events and to differences between urban, freeway, and rural driving segments. This analysis also determines the robustness of these metrics with respect to individual differences such as age and gender.

Due to simulator restarts, there were instances in which data could not be collected for the entire drive, indicated by cases where there were less than 108 data points. No efforts were made to replace the missing data. The tables (see Table 11 to Table 13) also show the lane deviation, average speed measures, and speed deviation measures, and lane deviation measures by BAC group. Figure 10 illustrates those differences across the events. Tables showing these measures by age group and gender are included in Appendix U.

7.4. Robustness of Metrics with Respect of Age, Gender, and Driver State

In contrast to the prior section where raw values were analyzed, this section will use composite score. These were chosen as the basis of the analysis because participants' performance and impairment may fluctuate across events resulting in impairment at the event level may be difficult to interpret. Composite scores for lane deviation, average speed, and speed deviation were examined to determine whether impairment was present across the entire drive. The composite scores were the t-scores ($M = 50$, $SD = 10$) based on the average of the z-scores of the measures across the events.

A 2 x 3 x 3 between-between-within ANOVA was performed on each of the three composite measures. Between-subjects independent measures were gender and age group (21-34, 38-51, 35-68). Within-subjects independent measure was (0.00%, 0.05%, and 0.10%). Because of multiple analyses, α was set to 0.01[4].

In computing the lane deviation composite score eleven participants were deleted from the analyses because of incomplete data. The analyses, therefore, were conducted with 97 subjects. The mean lane deviation composite scores by BAC, age group, and gender are shown in Appendix U. Mauchly's Test of Sphericity was not significant, indicating that no adjustment to the degrees of freedom was required.

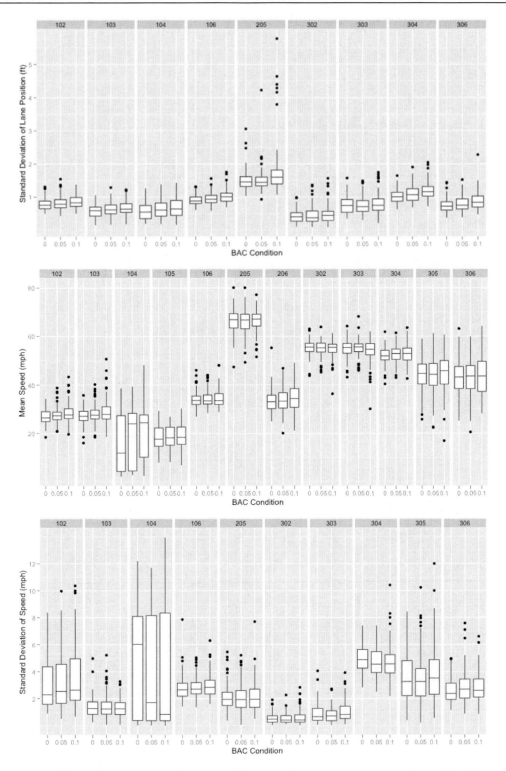

Figure 10. Effect of BAC on standard deviation of lane position, average speed, and speed deviation for Urban Drive (102), Green Light (103), Yellow Dilemma (104), Urban Curves (106), Interstate Curves (205), Lighted Rural (302), Transition to Dark Rural (303), Dark Rural (304), Gravel Transition (305), and Gravel Rural (306).

Table 11. Lane deviation (ft) by BAC group across events

Event	BAC Group											
	0.00% BAC			0.05% BAC			0.10% BAC			Total		
	M	N	SD	M	N	SD	M	N	SD	M	N	SD
Urban Drive (102)	.78	108	.17	.82	107	.21	.85	108	.18	.82	323	.19
Green Light (103)	.58	108	.19	.63	107	.22	.66	108	.23	.63	323	.21
Yellow Dilemma (104)	.59	107	.26	.63	108	.26	.70	105	.30	.64	320	.28
Urban Curves (106)	.90	108	.16	.94	108	.18	1.03	108	.21	.96	324	.19
Interstate Curves (205)	1.51	107	.30	1.51	108	.35	1.75	105	.75	1.59	320	.52
Lighted Rural (302)	.42	108	.19	.45	108	.26	.48	108	.27	.45	324	.24
Transition to Dark Rural (303)	.75	108	.25	.77	108	.28	.79	108	.30	.77	324	.28
Dark Rural (304)	1.02	108	.20	1.09	108	.23	1.20	106	.27	1.10	322	.25
Dark Rural Hairpin Curve (304)	.88	108	.27	.93	108	.32	.95	106	.34	.92	322	.31
Gravel Rural (306)	.75	108	.20	.81	108	.24	.87	106	.27	.81	322	.24

Note: BAC differences shown in bold are statistically significant at $p < 0.05$.

Table 12. Average speed (mph) by BAC group across events

Event	BAC Group											
	0.00% BAC			0.05% BAC			0.10% BAC			Total		
	M	N	SD	M	N	SD	M	N	SD	M	N	SD
Urban Drive (102)	**26.95**	**108**	**3.14**	**27.58**	**107**	**3.33**	**28.51**	**108**	**4.00**	**27.68**	**323**	**3.56**
Green Light (103)	**27.06**	**108**	**3.34**	**27.98**	**107**	**3.67**	**29.03**	**108**	**4.97**	**28.02**	**323**	**4.12**
Yellow Dilemma (104)	**15.79**	**108**	**11.82**	**18.17**	**108**	**11.41**	**20.42**	**108**	**10.81**	**18.13**	**324**	**11.48**
Left Turn (105)	18.35	108	4.65	18.48	108	4.52	18.99	108	4.64	18.61	324	4.60
Urban Curves (106)	33.92	108	3.21	34.21	108	3.22	34.44	108	3.66	34.19	324	3.37
Interstate Curves (205)	66.31	107	4.65	66.02	108	4.96	66.56	105	4.32	66.29	320	4.64
Exit Ramp (206)	33.26	108	4.94	33.72	108	5.14	34.41	108	5.24	33.80	324	5.11
Lighted Rural (302)	55.32	108	3.00	55.13	108	3.03	54.97	108	3.50	55.14	324	3.18
Transition to Dark Rural (303)	55.00	108	3.46	55.10	108	3.36	54.22	108	4.52	54.77	324	3.82
Dark Rural (304)	52.03	108	3.21	52.58	108	3.39	52.83	107	3.62	52.48	323	3.41
Dark Rural Hairpin Curve (304)	44.86	108	4.46	45.78	108	4.09	45.50	107	4.69	45.38	323	4.42
Gravel Transition (305)	44.40	108	6.52	44.09	108	6.74	44.82	107	7.43	44.44	323	6.89
Gravel Rural (306)	43.07	108	7.06	43.20	108	7.32	43.97	106	8.14	43.41	322	7.50

Note: BAC differences shown in bold are statistically significant at $p < 0.05$.

Table 13. Speed deviation (mph) by BAC group across events

Event	BAC Group											
	0.00% BAC			0.05% BAC			0.10% BAC			Total		
	M	*N*	*SD*	*M*	*N*	*SD*	*M*	*N*	*SD*	*M*	*N*	*SD*
Urban Drive (102)	3.03	108	1.85	3.25	107	2.00	3.52	108	2.24	3.27	323	2.04
Green Light (103)	1.33	108	.73	1.42	107	.84	1.34	108	.67	1.36	323	.75
Yellow Dilemma (104)	4.52	108	3.88	4.21	108	4.07	4.03	108	4.38	4.25	324	4.11
Urban Curves (106)	2.75	108	.86	2.82	108	.77	2.93	108	.85	2.84	324	.83
Interstate Curves (205)	2.08	107	.94	2.08	108	1.01	2.17	105	1.11	2.11	320	1.02
Lighted Rural (302)	.53	108	.36	.51	108	.37	.60	108	.53	.55	324	.43
Transition to Dark Rural (303)	**.90**	**108**	**.72**	**.80**	**108**	**.55**	**1.04**	**108**	**.75**	**.91**	**324**	**.68**
Dark Rural (304)	4.91	108	.95	4.66	108	1.05	4.69	107	1.26	4.76	323	1.10
Dark Rural Hairpin Curve (304)	2.73	108	1.20	2.61	108	1.15	2.55	107	1.22	2.63	323	1.19
Gravel Transition (305)	3.51	108	1.80	3.45	108	1.82	3.76	107	1.95	3.57	323	1.85
Gravel Rural (306)	2.60	108	.89	2.81	108	1.22	2.81	106	1.10	2.74	322	1.08

Note: BAC differences shown in bold are statistically significant at $p < 0.05$.

Of the within-subjects effects, the only statistically significant effect was the main effect of BAC, F (2, 182) = 34.82, $p < 0.001$, partial $\eta^2 = 0.28$. As shown in Figure 11, lane deviation composite scores increased as a function of BAC, producing a statistically significant linear trend, F (1, 91) = 60.78, $p < 0.001$, partial $\eta^2 = 0.28$. The quadratic trend was not significant, F (1, 91) = 1.49, $p > 0.05$, partial $\eta^2 = 0.02$. No between-subjects effect or interactive effects were statistically significant.

The average speed deviation composite scores by BAC, age group, and gender are shown Appendix U. Mauchly's Test of Sphericity was statistically significant, and the Greenhouse-Geisser adjustment was used to adjust the degrees of freedom. Of the within-subjects effects, the only statistically significant effect was the main effect of BAC, F (1.89, 192.27) = 6.59, $p < 0.01$, partial $\eta^2 = 0.06$. As shown in Figure 12, average speed composite scores decreased as a function of BAC, producing a statistically significant linear trend, F (1, 102) = 11.70, $p < 0.01$, partial $\eta^2 = 0.10$. The quadratic trend was not significant, F (1, 102) = 0.00, $p > .05$, partial $\eta^2 = 0.00$. No interactive effects were statistically significant.

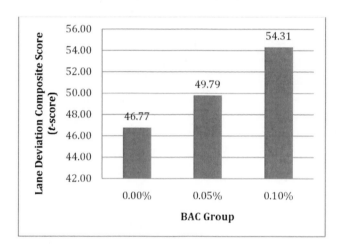

Figure 11. Lane deviation as a function of BAC group.

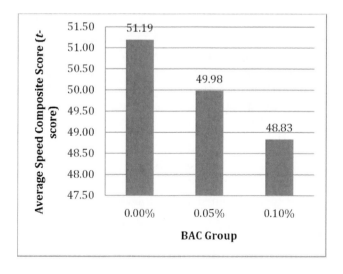

Figure 12. Average speed as a function of BAC group.

Table 14. Measured BAC levels associated with the two BAC classes

Test Time	BAC < 0.08% (N = 224)			BAC ≥ 0.08% (N = 89)			Total (N = 313)		
	M	*SD*	Median	*M*	*SD*	Median	*M*	*SD*	Median
Pre-drive	.028	.027	.042	.101	.004	.103	.049	.041	.054
Post-drive	.023	.023	.032	.091	.006	.092	.042	.037	.043
Average	.025	.025	.037	.096	.004	.097	.045	.038	.048

Of the within-subjects effects, the only statistically significant effect was the main effect of BAC, F (2, 102) = 8.81, $p < 0.001$, partial $\eta^2 = 0.15$. Of the between-subjects effects, the only statistically significant effect was the main effect of Gender, F (1, 102) = 8.47, $p < 0.01$, partial $\eta^2 = 0.08$. Speed deviation was greater among males (52.23) than among females (47.77). The mean speed deviation composite scores by BAC, age group, and gender are shown in Appendix U. Mauchly's Test of Sphericity was not significant, indicating that no adjustment to the degrees of freedom was required. No interactive effects were statistically significant.

The analyses were also performed with sleepiness as a covariate. There was a within-subjects interaction of BAC and sleepiness for lane deviation, F (2, 180) = 5.97, $p < 0.01$, partial $\eta^2 = 0.12$; and a within-subject interaction of BAC and sleepiness for average speed, F (1.89, 190.57) = 4.21, $p < 0.05$, partial $\eta^2 = 0.08$. Thus, sleepiness was associated with some of the effects of BAC, which is not surprising given that BAC is positively correlated with sleepiness, $r = 0.25$, $p < 0.001$.

7.5. Conclusion

Analysis of common driving metrics demonstrates the sensitivity of the drive to alcohol impairment. As expected, lane position variation was particularly sensitive and speed variation was less so. Increasing BAC levels generally affected driving performance in an orderly manner—higher BAC levels led to a linear decrease in performance. Drivers' response to the events was generally robust and unaffected by repeating the drives, with the variation between events being much greater than the variation between drives. The metrics were also robust to the effects of age and gender. No interactions affected driver performance and only the composite speed score was influenced by the main effects of age and gender. Alcohol levels did not interact with age, gender, and roadway situation, which might have otherwise undermined the association of driving metrics and alcohol impairment.

8. ALGORITHM DEVELOPMENT AND EVALUATION

The primary objectives for algorithm development and evaluation include:

- Develop algorithms to detect alcohol-related impairment based on behavioral signatures that vehicle-based sensors can measure
- Compare sensitivity, robustness, and timeliness of metrics and algorithms.

This section addresses these objectives by first describing the performance of a logistic regression algorithm that builds directly on an analysis of simple measures of driving performance—lane position variability, mean speed, and speed variability. To go beyond these three simple indicators of driver impairment, a decision tree algorithm fit to individual events and to the urban, freeway, and rural segments identifies behavioral signatures of alcohol impairment. These signatures provide a detailed description of alcohol impairment that supports more accurate detection that the three-variable logistic regression. The final sections of the section assess algorithm sensitivity, robustness, timeliness, and bias defined as:

Sensitivity—The number of correctly classified cases—true positives and true negatives
Robustness—Vulnerability to generalization error, context, or available data
Timeliness—The amount of data aggregated over time to produce an accurate classification
Bias—The tendency to favor detecting impairment at the expense of incorrectly identifying impairment when there is none.

8.1. Data and General Methods of Algorithm Development and Evaluation

The objective of the following analyses was to determine whether it is possible to distinguish between drivers with BACs at and above 0.08% and those below 0.08%. To that end, a new variable was created (BAC Status) by dichotomizing the pre- and post-drive BACs as either both being less than 0.08% or both being at or above 0.08%. The dichotomization produced 313 valid cases. Eleven cases were eliminated because the pre- and post-drive BACs were not on the same side of the 0.08% cutoff. BAC status characteristics are shown in Table 14. The median BAC for the low BAC status condition (BAC < 0.08%) was 0.037%. The median BAC for the high BAC status condition (BAC ≥ 0.08%) was 0.097%. The median differences between the conditions were 0.06%.

Three general algorithms were developed. The first was based on logistic regression and was fit using a standard least squares regression approach using the entire dataset. The two other approaches to algorithm development used support vector machines (SVMs) and decision trees, which can often outperform linear combinations of the features (Liang, et al., 2007b). Originally developed by Vapnik (1995), SVMs have several advantages over approaches that make assumptions of linearity and normality. The SVM approach identifies a hyperplane that separates instances with different BAC levels (Saarikoski, 2008). SVMs are particularly well-suited to extract information from noisy data (Byun & Lee, 2002) and avoid overfitting by minimizing the upper bound of the generalization error (Amari & Wu, 1999). A decision tree approach, C4.5, classifies data by creating a tree that divides the data using the gini index, which weights feature influence in a linear fashion (Lim, Loh, & Shih, 2000; Quinlan, 1996). Adaptive boosting (AdaBoost) sequentially fits a series of classification algorithms, with greater emphasis on previously misclassified instances. It then combines the output of the classification algorithms by adjusting the importance of each classifier based on its error rate (Freund & Schapire, 1996). This approach is particularly valuable where a single decision tree or SVM cannot capture the complexity of the underlying relationships. Adaptive boosting was applied to both the Decision Tree and SVM, but not the logistic regression.

Three criteria are used throughout to assess algorithm sensitivity: accuracy, positive predictive performance (PPP), and area under curve (AUC). Accuracy measures the percent of cases that were correctly classified, and PPP measures the degree to which those drivers that were judged to have high BAC levels actually had high BAC levels. Performance measures such as correct detection or overall accuracy fail to provide a complete description of algorithm performance because they do not account for the baseline frequency of impairment nor differences in the decision criterion. An algorithm can correctly identify all instances of impairment simply by setting a very low decision criterion, but such an algorithm would misclassify all cases where there was no impairment. The signal detection parameter, d', avoids these problems, but its underlying assumptions include symmetry of signal and noise distributions, which are often violated. AUC is a nonparametric version of d' and represents the area under the receiver operator curve, which provides a robust performance measure that does not depend on the assumptions underlying d'. Perfect classification performance is indicated by an AUC of 1.0, and chance performance is indicated by 0.50. AUC is an unbiased measure of algorithm performance, but accuracy and PPP are more easily interpreted, so all three are used in describing the algorithms.

8.2. Logistic Regression Algorithm and Basic Driving Performance

A sequential logistic regression was performed to assign each case to one of the two BAC categories (BAC < 0.08% or BAC ≥ 0.08%), first using 11 speed deviation measures, then after adding 13 average speed measures, and then after adding 10 lane deviation measures. Ten cases had missing data and were deleted, leaving 303 cases for the logistic regression.

A test of the speed deviation predictors against a constant-only model was not statistically significant, $\chi^2(11, N = 303) = 8.26$, $p > .05$. Adding 13 average speed predictors produced a statistically significant improvement, $\chi^2(13, N = 303) = 32.74$, $p < .01$. Adding the lane deviation predictors was also statistically significant, $\chi^2(10, N = 303) = 54.38$, $p < .001$. These results are summarized in Table 15. Only measures of average speed and lane deviation, therefore, were useful in predicting BAC status. Figure 13 shows the increase of overall classification accuracy as a function of the events composing the drive.

Table 15. Logistic regression for BAC status as a function of speed deviation, average speed, and lane deviation across segments

Measures, variables, and regression parameters						95% Confidence Interval for Odds Ratio	
Measure	Variable	B	Wald Test	p	Odds Ratio	Lower	Upper
Speed Deviation	Urban Drive (102)	.254	4.460	.035	1.290	1.018	1.633
	Green Light (103)	-.323	1.305	.253	.724	.416	1.260
	Yellow Dilemma (104)	.301	8.559	.003	1.351	1.104	1.653

	Measures, variables, and regression parameters					95% Confidence Interval for Odds Ratio	
	Urban Curves (106)	-.167	.561	.454	.846	.546	1.311
	Interstate Curves (205)	-.195	.993	.319	.823	.561	1.207
	Lighted Rural (302)	.008	.000	.985	1.008	.455	2.229
	Transition to Dark Rural (303)	.150	.244	.622	1.162	.640	2.109
	Dark Rural (304)	-.101	.245	.620	.904	.606	1.348
	Dark Rural Hairpin Curve (304)	.008	.003	.958	1.008	.748	1.359
	Gravel Transition (305)	.123	1.231	.267	1.131	.910	1.405
	Gravel Rural (306)	-.141	.650	.420	.869	.617	1.223
Average Speed	Urban Drive (102)	.067	.337	.561	1.069	.853	1.339
	Green Light (103)	.075	.486	.486	1.078	.873	1.331
	Yellow Dilemma (104)	.106	7.072	.008	1.111	1.028	1.201
	Left Turn (105)	.093	2.155	.142	1.097	.969	1.242
	Urban Curves (106)	-.285	8.163	.004	.752	.618	.914
	Interstate Curves (205)	-.012	.042	.838	.988	.881	1.108
	Exit Ramp (206)	.073	2.178	.140	1.075	.977	1.184
	Lighted Rural (302)	.002	.000	.985	1.002	.791	1.270
	Transition to Dark Rural (303)	-.186	2.320	.128	.831	.654	1.055
	Dark Rural (304)	-.024	.049	.824	.976	.786	1.211
	Dark Rural Hairpin Curve (304)	-.072	1.304	.253	.930	.821	1.053
	Gravel Transition (305)	.053	1.316	.251	1.054	.963	1.154
	Gravel Rural (306)	-.009	.050	.823	.992	.920	1.069
	Urban Drive (102)	-2.196	4.190	.041	.111	.014	.911

Table 15. (Continued)

Measures, variables, and regression parameters						95% Confidence Interval for Odds Ratio	
Lane Deviation	Green Light (103)	.426	.234	.628	1.532	.272	8.609
	Yellow Dilemma (104)	1.899	4.670	.031	6.680	1.193	37.394
	Urban Curves (106)	3.211	6.885	.009	24.796	2.253	272.873
	Interstate Curves (205)	.758	3.828	.050	2.135	.999	4.564
	Lighted Rural (302)	-.451	.350	.554	.637	.143	2.836
	Transition to Dark Rural (303)	-.590	.808	.369	.554	.153	2.007
	Dark Rural (304)	2.638	7.582	.006	13.989	2.139	91.492
	Dark Rural Hairpin Curve (304)	-.337	.270	.603	.714	.200	2.546
	Gravel Transition (306)	1.369	3.436	.064	3.933	.924	16.734
	Constant	4.285	1.067	.302	72.613		

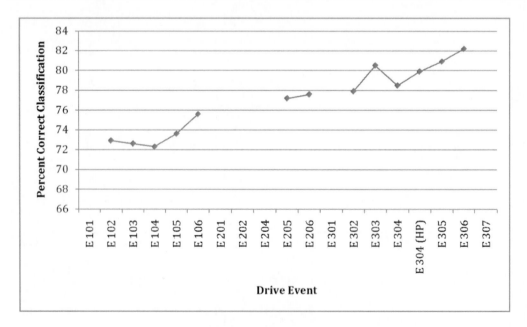

Figure 13. Correct classification for the cumulative logistic regression algorithm.

Overall classification performance was acceptable. The 11 speed deviation variables and the 13 average speed variables produced a correct classification of 95% for low BAC status and 21% for high BAC status (PPP), with an overall correct classification of 74%.

The addition of the lane deviation variables resulted in a correct classification of 95% for the low BAC status and a correct classification of 49% for the high BAC status, with an overall correct classification of 82%. At the last event, the variance in BAC status accounted for by the algorithm was moderate, with Nagelkerke $R^2 = .39$. The AUC metric for this analysis was .720 (CI=0.073). Table 15 shows the regression coefficients, Wald statistic, odds ratios, and the 95% confidence interval for odds ratios for the 34 predictors of BAC status, based on all events.

It could be argued that logistic regression is not the correct analytic tool for these data. Logistic regression is a between-subjects strategy, which assumes the cases are unrelated to each other. In the current study, however, each subject was tested three times. The correlations of the within- subjects runs are not accounted for in the logistic regression model. To address these concerns, separate analyses were conducted, adjusting for individual differences such as age, gender, height, weight, and learning. The results of those analyses were very similar to those presented above and are not presented here. This analysis provides a baseline and point of comparison for the more complex algorithms.

8.3. Signatures of Alcohol Impairment

The diversity of driving situations and associated driver responses might provide behavioral signatures of impairment that are more sensitive than the simple measures of lane position variability, mean speed, and speed variability. To assess this possibility, a decision tree algorithm was fit to each event using a diverse set of variables, with the results shown in Table 16. These variables reflect the range of cues police officers use to detect impaired drivers discussed earlier and cataloged in Table 3.

The variables in Table 16 and throughout the document are abbreviated with a consistent notation. The first part of the variable indicates the measure, such as lane position (lp_), normalized lane position (lpn_), speed (sp_), and acceleration (acc_). The average, minimum, maximum, and standard deviation are shown as (_avg , _min, _max, _sd), and the initial and final values for an event are shown as _init and _end. Appendix B contains a complete list and detailed definition of each variable. Some of these variables, such as turn signal use, are relevant to only a few events, but others, such as minimum speed, are applicable to many.

Table 16 shows the variables indicative of alcohol impairment and how well a decision tree analysis can combine them to separate drivers with high and low levels of BAC. The decision tree analysis identifies a set of variables and their associated levels that best separate one condition from another. AUC values show that most events contain diagnostic information, with the exception of Merge On (202), which has an AUC of 0.50. Some events are highly sensitive to alcohol impairment, with AUCs exceeding 0.75. Dark Rural (304) is the most sensitive event with an AUC of 0.84. Similar to the other sensitive events in Table 16, this event is relatively long and places substantial demands on the driver. Dark Rural requires drivers to negotiate an unlit rural road with a sharp curve. Urban Drive (102), while very different in the details, is also quite sensitive, with an AUC of 0.78. The Urban Drive places substantial demands on the driver to scan the environment, maintain position within a relatively narrow lane, and monitor surrounding vehicles and pedestrians. This analysis shows that the behavioral signatures associated with various driving situations differ substantially in their sensitivity to alcohol. For six events, the sensitivity of the decision tree algorithm exceeded that of the logistic regression even when the logistic regression included data from entire drive.

The greater sensitivity of the decision tree algorithm relative to the logistic regression reflects, in part, different variables. Although some variables are shared with the logistic regression analysis, many are notably different. Urban Drive (102), for instance, includes the speed and lane variability measures used in the logistic regression, as well as minimum speed and standard deviation of normalized lane position). The standard deviation of normalized lane position (lpn_sd) is the variability relative to the lane center, rather than the variability relative to the mean lane position (lp_sd).

Table 16. Decision tree algorithm applied to each event of the drive

Segment	Event Name (number)	Description (variables)	Approximate Duration (seconds)	AUC (Accuracy)
Urban	Pull Out (101)	Pull out of parallel parking spot into traffic (gap_taken_t, turn_signal)	30	.60
	Urban Drive (102)	Drive on a narrow 2-lane road with traffic and parked vehicle (sp_avg, lpn_sd, lp_sd, sp_end, sp_min, sp_sd)	45	.78
	Green Light (103)	Navigate green traffic light on urban2-lane road with parked vehicles along the road, oncoming traffic, traffic behind driver (sp_min, spn_avg sp_min, brake_press)	30	.66
	Yellow Dilemma (104)	Navigate yellow light dilemma on urban two-lane with parked vehicle, oncoming traffic, traffic behind driver (lpn_sd, sp_max, spn_avg, spn_avg spn_avg)	75	.64
	Left Turn (105)	Left turn at signalized intersection (no green arrow, no dedicated turn lane), oncoming traffic, variety of gaps (sp_max, sp_avg, sp_min)	80	.65
	Urban Curves (106)	Three curve segments of mixed radius of curvature (lpn_sd, lp_avg, spn_sd, sp_sd)	180	.80
Freeway	Turn On Ramp (201)	Turn right onto interstate on-ramp (accel_release, sp_end, acc_avg, sp_init)	30	.61

Segment	Event Name (number)	Description (variables)	Approximate Duration (seconds)	AUC (Accuracy)
	Merge On (202)	Merge onto interstate (acc_avg)	50	.50
	Interstate Curves (205)	Navigate three curves on interstate (lp_sd, sp_avg, sp_sd)	185	.65
	Exit Ramp (206)	Take exit ramp off interstate (acc_avg, sp_avg)	30	.65
Rural	Turn Off Ramp (301)	Turn right from ramp onto rural two- lane road (stop_pos, sp_end, acc_done_d, spe_end, acc_end, acc_done)	30	.73
	Lighted Rural (302)	Lighted two-lane rural road, 55 mph (lp_avg, spn_sd, lp_sd)	90	.68
	Transition to Dark Rural (303)	Straight roadway that transitions between lighted and unlit (sp_sd, sp_avg, lp_sd, lpn_sd, sp_init)	20	.65
	Dark Rural (304)	Unlit straight and curved road, segments, center and road edge marking are faded and the roadsurface is grayish. A hairpin turn and a vertical curve (lp_sd, lp_sd_hp, lp_avg, sp_min, lpn_sd_hp, sp_sd)	300	.84
	Gravel Transition (305)	Transition to gravel surface on straight road (steer_sd, sp_avg, sp_sd, sp_init)	30	.73
	Gravel Rural (306)	Gravel road (straight and curves) (lp_sd, lpn_sd, sp_end, sp_init, spn_sd)	90	.76
	Driveway (307)	Pull into driveway with gravel (steer_max, sp_end, sp_init, turn_signal)	30	.63

Figure 14 shows the decision trees associated with Urban Drive (102). Rounded rectangles show variables used to divide instances into high and low BAC. Numbers on the lines indicate criteria for the divisions. The rectangles represent leafs of the decision tree—the classification outcome associated with the criteria leading to the leaf. The label indicates whether the instances associated with the leaf are identified as high (TRUE) or low (FALSE) BAC levels. The width of the bar of each leaf indicates the number of instances that correspond to the conditions associated with the leaf. The clear and filled components of this bar indicate the proportion of high and low BAC levels: Clear corresponds to low and filled to high. A perfectly accurate decision tree would have bars that are either entirely clear or entirely filled.

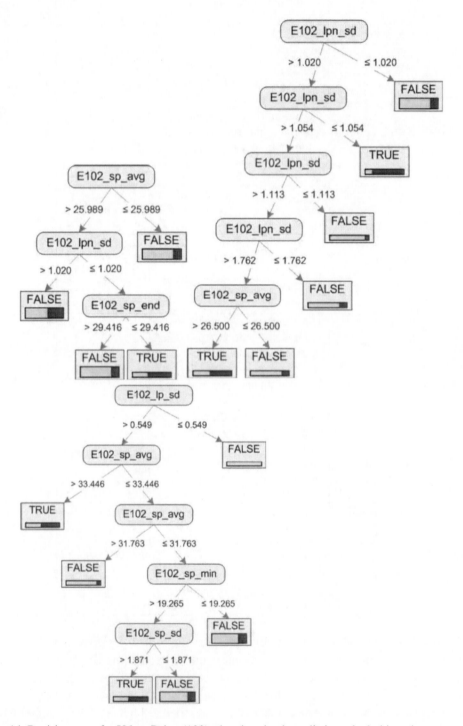

Figure 14. Decision trees for Urban Drive (102), showing the three distinct alcohol impairment signatures.

Figure 15. Decision trees for Urban Curves (106), showing the three distinct signatures of impairment.

The trees are developed sequentially according to the ADAboost process, with the second tree being fit to those cases that the first misclassified. Because each tree represents a solution for those cases that were not fit by the others, each reveals a different behavioral signature of impairment. For Urban Drive (102), the first decision tree shows that a combination of speed and lane variability identifies alcohol impairment—those driving faster and with greater lane position variability tend to be those with high BAC levels. The second tree shows lane position as the dominant differentiating factor, followed by speed variability. The third tree uses a combination of lane position variability relative to the mean lane position, as opposed to the centerline, as in other two trees. These figures demonstrate that alcohol impairment might be detected most efficiently by a collection of variables combined in a non-linear fashion.

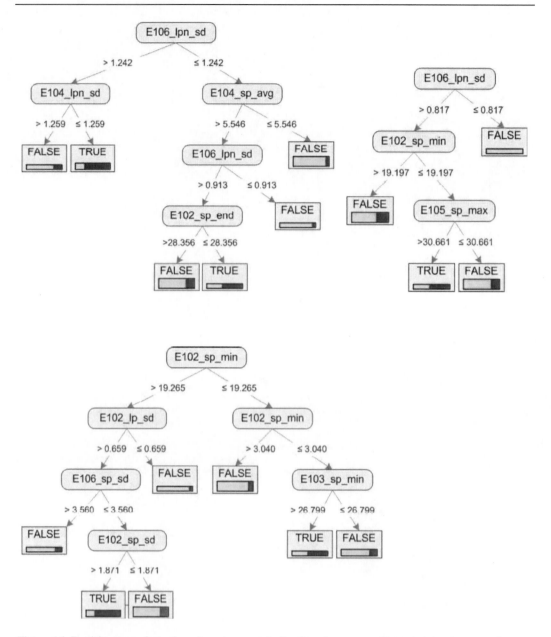

Figure 16. Decision trees from the urban segment, indicating signatures of impairment that span the Urban Drive (102), Green Light (103), Yellow Dilemma (104), Left Turn (105), and Urban Curves (106) events.

The decision trees associated with the Urban Curves (106) in Figure 15 show signatures of alcohol impairment that share many features with those of Urban Drive (102). In both, normalized lane position variability plays a dominant role. The average lane position is also part of the behavioral signature for this event, reflecting a tendency of impaired drivers to straddle the lane in curves, but not in the dense urban environment of Urban Drive (102). Another notable difference is the role of speed variability, which was not a factor in Urban Drive (102) or the logistic regression, but emerges as in important feature in Urban Curves

(106). This may reflect the diminished ability of drivers with high BAC to manage the dual tasks of lane keeping and speed control that are both demanding on the rural curves. The most striking difference between these two events is the decision tree complexity: the trees for Urban Curves (106) have 19 nodes in total, compared to 12 for Urban Drive (102). This and the variety of variables associated with the decision trees for each event support two important conclusions: alcohol has a clear and consistent effect on lane-keeping performance, and lane position variability combines with speed and other variables in a way that depends on the particular driving context.

A notable feature of both of both figures is the occasional occurrence of seeming arbitrary levels of a variable that are associated with impairment, but other levels are not—the standard deviation of lane position in the second decision tree in Urban Drive (102) as an example. Some of these reflect the complex relationship between variables, such as when impaired drivers drive more slowly, which tends to reduce their lane position variability relative to unimpaired drivers who adopt a higher speed. Other instances might represent cases where the decision trees over fit the data, so the accuracy might be lower if the decision trees were applied to data not used in their construction; this is a topic that we will return to in the analysis of algorithm robustness.

Building decision trees with data from the urban, freeway, and rural segments highlights impairment signatures that cut across several events. Figure 16 shows how lane position dominates, but that factors such as average speed combine to identify a set of unimpaired drivers, such as those that have a low standard deviation of normalized lane position on Urban Curves (106) (E106_lpn_sd) and who stopped at the yellow light dilemma (E104_sp_avg). In the second and third decision trees, the minimum speed in Urban Drive (102) (E102_sp_min) is combined with speed metrics from other events to indicate aberrant speed control in drivers whose lane position variability did not indicate high BAC.

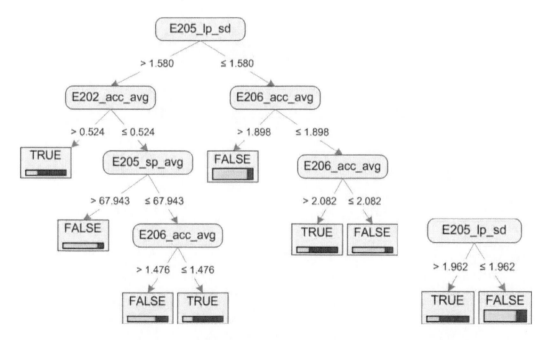

Figure 17. Decision trees from the freeway segment, indicating signatures of impairment that span the Merge On (202), Interstate Curves (205) and Exit Ramp (206) events.

Figure 17 shows the decision trees for the freeway segment. In the first decision tree, the degree of abrupt acceleration and deceleration helps identify impaired drivers, but the standard deviation of lane position during the curves on the interstate dominate both decision trees. High BAC leads to poor lane keeping, abrupt acceleration, and slower speeds.

Figure 18 shows the decision trees and associated signatures of impairment for the rural segment. Similar to the other segments, the first and fourth decision trees show that lane position variability strongly differentiates between impaired and unimpaired drivers. The other decision trees show that poor speed control differentiates between high and low BAC.

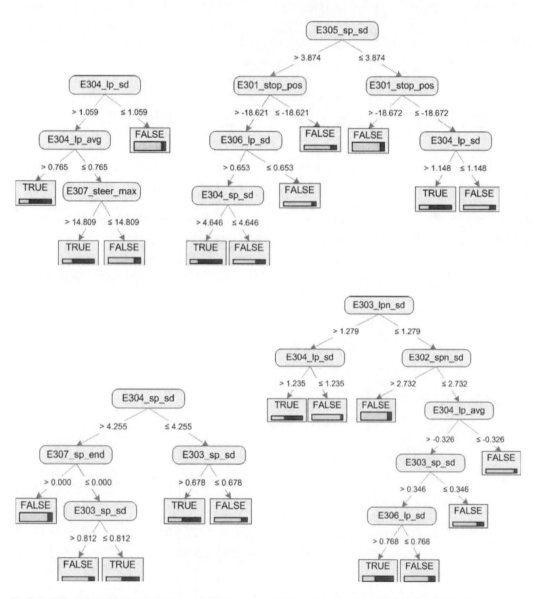

Figure 18. Decision trees from the rural segment, indicating signatures of impairment that span the Turn Off Ramp (301), Light Rural (302), Transition to Dark Rural (303), Dark Rural (304), Gravel Transition (305), Gravel Rural (306) and Driveway (307) events.

The classification performance of the decision trees that combine data over the urban, freeway, and rural segments substantially exceeds that of logistic regression, which included data from the entire drive. The logistic regression algorithm produced an AUC value of .720, compared to .824 for the decision tree algorithm applied only to the urban segment. The decision trees for the freeway and rural segments were similarly diagnostic with AUCs of .727 and .851, respectively. Positive predictive performance measures the degree to which those drivers that were judged to have high BAC levels actually had high BAC levels. The urban segment produced a PPP of 89.5%, and the freeway and rural segments produced PPP of 65.2% and 74.4%, respectively. The overall accuracy was 81.2%, 78.3%, and 77.6% for the urban, freeway, and rural segments.

The decision tree algorithm reveals important signatures of impairment not apparent in the logistic regression algorithm. Consistent with the logistic regression analysis, lane position variability emerged as a dominant indicator of impairment; however, several context-specific variables, such as maximum acceleration and minimum speed, proved to be indicative of high BAC levels, but only for specific events. These signatures of impairment strongly suggest that impairment-detection algorithms should consider the driving context.

Algorithm development ultimately aims not to identify signatures of impairment, but to identify a sensitive, robust, and timely indicator of impairment. With a mean accuracy of 79.0% for the three segments of the drive, the decision tree algorithm confirms that it is possible to create a diagnostic algorithm that is not tailored to an individual driver. The logistic regression algorithm achieved an accuracy of only 74.4% by combining information across the entire drive, achieving maximum sensitivity after approximately 25 minutes of driving. The decision tree algorithm is much more timely, identifying impairment in these situations with greater precision after only approximately 8 minutes of driving. This analysis also shows that timely impairment detection depends critically on the driving context: specific variables that reflect the demands of challenging events result in a much more timely impairment detection than generic variables.

8.4. Assessing and Enhancing Algorithm Sensitivity, Robustness, and Timeliness

The following section further explores the degree to which the decision tree algorithm, by assessing how robust it is to generalization and whether other algorithms, such as one based on support vector machines, can provide a more timely and sensitive assessment of impairment. More generally, the following sections assess the degree to which decision trees and support vector machine algorithms are robust, and the degree to which they can be made more robust, timely, and sensitive. For these analyses, performance of the standardized field sobriety test provides a point of comparison.

8.4.1. Standardized field sobriety test as a baseline for algorithm sensitivity
The common application of the SFST in assessing alcohol impairment of drivers makes it a useful benchmark for assessing algorithm sensitivity—i.e., a point of comparison to see if algorithms are as good as SFST in determining impairment. Ideally, algorithms using vehicle-based sensors would exceed the capacity of the SFST to discriminate between BAC levels.

This section applies three algorithms to the SFST to identify a level of accuracy that can be used to assess behavior-based algorithms later in the report.

Several studies of the SFST have demonstrated that a battery of simple visual motor tasks can discriminate between people with BAC levels above or below 0.10 with an accuracy of 83% (Burns & Moskowitz, 1977). This accuracy was based on data from five groups: Group 1 had an average BAC of 0.000% ($N = 79$), Group 2 had an average BAC of 0.041% ($N = 20$), Group 3 had an average BAC of 0.073% ($N = 20$), Group 4 had an average BAC of 0.120% ($N = 48$), and Group 5 had an average BAC of 0.156% ($N = 16$). The SFST battery was later reduced to a battery of three subtests: horizontal gaze nystagmus, walk and turn, and one leg stand. This reduced battery discriminated between those with a BAC of 0.10% or higher with an accuracy of 81% (Tharp Burns Moskowitz, 1981). These results were obtained in the laboratory, with relatively inexperienced testers.

More recent validation studies with more experienced testers indicate that overall accuracy is much higher in the field. In a validation study using 0.08% as the BAC criterion, officers were able to correctly identify 96% of drivers with BAC $\geq$ 0.08%, 93% of drivers with BAC < 0.08%, for an overall accuracy of 93% (Burns & Dioquino, 1997). These results are similar to those of a more recent study (Stuster, 2006). These results suggest the SFST can be very sensitive to BAC; however, police officers applying the SFST in the field studies were not blind to driving behavior or driver state, such as aberrant driving, drivers slurring words, and the smell of alcohol in the vehicle. Such cues may have contributed to the high classification accuracy.

This study placed drivers into three conditions (0.0%, 0.05%, and 0.10% BAC) with the aim of producing declining levels of BAC while the participants were in the simulator. The mean BACs for the three experimental conditions were 0.000% ($N = 108$, $SD = 0.000%$), 0.047% ($N = 108$, $SD = 0.005%$), and 0.093% ($N = 108$, $SD = 0.008%$), respectively. None of the drivers reached the higher BAC levels observed in the previous studies, making the discrimination task of the algorithms more difficult. Previous investigations using the SFST to classify BAC levels derived and applied decision criteria to the same sample, potentially inflating detection accuracy. A more conservative approach involves cross validation that derives decision criteria from one sample and applies them to another sample.

To identify SFST performance as a baseline for comparing the algorithm, two situations were considered: the first discriminating between the experimental conditions of 0.00% BAC and 0.10% BAC and the second between drivers above and below 0.08% BAC. Based on previous studies, one might expect greater accuracy in discriminating between people with a large difference in BAC. This analysis is based on the SFST administered immediately after the drive and scored according to the protocol in Appendix Q. Analysis of the SFST considers three techniques for discriminating between BAC levels: logistic regression, decision tree, and support vector machine.

Classification accuracy was consistent with previous studies—classification accuracy exceeded 82% for all three algorithms, with the decision tree being most accurate (84.7), followed by SVM (82.3) and logistic regression (82.0). Not surprisingly, the performance discriminating between BAC levels above and below 0.08 was somewhat worse than between the more extreme range defined by the experimental conditions of 0.00% and 0.10% BAC. Table 17 shows the accuracy ranges from approximately 80.5 to 82.5%. The failure to perfectly discriminate between BACs is not surprising given the overlap in SFST scores across BAC levels, as shown in Figure 19.

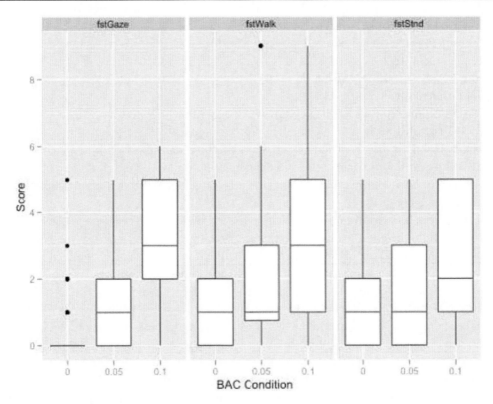

Figure 19. SFST scores show considerable overlap across BAC conditions.

Table 17. Performance of three algorithms classifying drivers with BAC above and below 0.08% using the SFST, with confidence intervals in the parentheses

	Accuracy	AUC	PPP
Decision tree	81.8 (5.9)	.76 (0.087)	78.4 (15.5)
SVM	80.5 (6.9)	.81 (0.072)	75.6 (17.9)
Logistic regression	82.5 (5.5)	.80 (0.062)	75.9 (13.6)

Algorithms for combining elements of the SFST correctly predicted BAC levels with similar accuracy as that found in previous laboratory studies, but with lower accuracy than recent field studies. Reasons for the lower accuracy include less experienced administrators and a cross- validation statistical analysis that produces accuracy estimates that may be more representative predictions using data that are not included in estimating algorithm parameters. Another factor that contributes to algorithm performance is the BAC level being discriminated. The highest average BAC in the current study was 0.108%. This was dictated by practical and ethical reasons. The range of BACs in the current study, therefore, is narrower than any previous SFST studies, either in the laboratory or in the field. The importance of BAC range is demonstrated by the four to five percent greater accuracy when discriminating between placebo and the 0.10% BAC condition, compared to discriminating between those above and below 0.08% BAC. In addition, police officers in the recent field studies likely benefitted from a range of cues beyond those in the SFST, which were not available to those performing the SFST in this study (Rubenzer & Stevenson, 2010).

8.4.2. Algorithm sensitivity and robustness

Robustness is the degree to which algorithm performance depends on factors unrelated to driver impairment, such as road type and individual differences. This analysis considers three elements of robustness: generalization error, dependence on individual differences, and the effect of different road types.

Generalization error is the degree to which algorithm sensitivity declines when it is applied to data that were not included in its development. An algorithm that identifies impairment based on the idiosyncratic behaviors of particular drivers in a particular driving context will perform well when applied to the data used to develop the algorithm, but poorly when applied to data not used in its development. A large generalization error indicates over fitting and poor robustness.

Sensitivity of an algorithm assessed with cross validation was compared to sensitivity of an algorithm fit to the entire data set to estimate generalization error. In cross validation, one data set is withheld and the algorithm is trained on the remaining data set and then tested on the withheld data. This study used a 10-fold cross validation, in which 10 datasets were created, withholding for testing a stratified sample of 10% of the data. This produces 10 estimates of algorithm performance, which are averaged to assess the algorithm performance. Results are not typically sensitive to the number of folds in the cross validation process, but a 10-fold is most commonly used (Efron & Gong, 1983; Feng, et al., 2008).

Table 18 shows two aspects of algorithm robustness: generalization error and sensitivity to the driving context. Cross validation reveals substantial generalization error within all three roadway segments and relatively robust performance between the three segments. Accuracy was approximately 10% lower for all road segments with cross validation. Likewise AUC and PPP were substantially lower, with AUC dropping from by approximately 0.15. In contrast, the algorithm was quite robust to the differences between segments. AUC for the cross validation ranged from 0.63 to 0.65 across segments, almost an order of magnitude less than the generalization error. This difference reflects the tailoring of the algorithm to the event— signatures of alcohol impairment were derived for measures from each event rather than a single criterion applied to all events.

Table 18. Cross validation to assess robustness associated generalization error and driving context

	Accuracy	AUC	PPP
Urban segment			
Decision tree fit to all data	81.2	0.82	89.5
Decision tree cross validation	70.3 (3.9)	0.65 (0.061)	50.2 (18.7)
Freeway			
Decision tree fit to all data	78.3	0.73	65.2%
Decision tree cross validation	68.7 (5.5)	0.63 (0.091)	43.1 (24.9)
Rural			
Decision tree fit to all data	77.6	0.85	74.4%
Decision tree cross validation	73.2 (6.5)	0.65 (0.11)	55.2 (25.1)
Field sobriety test cross validation	81.8 (5.9)	0.76 (0.087)	78.4 (15.5)

95% confidence interval in parentheses.

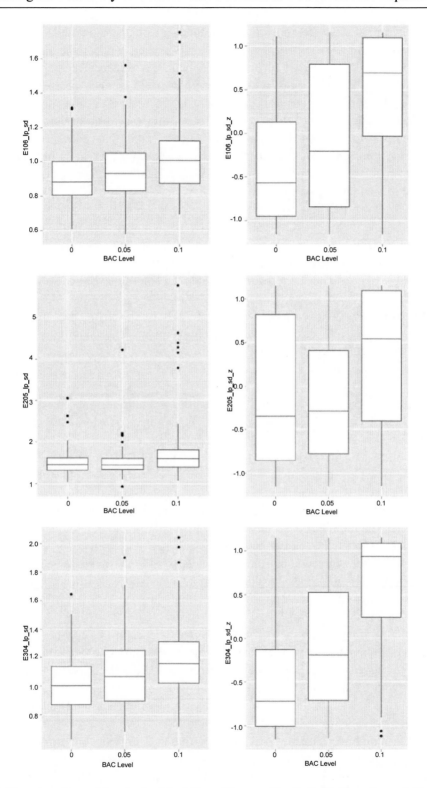

Figure 20. The z-transform of the standard deviation of lane position individualizes the model input and separates BAC levels more clearly.

Individualization describes how the algorithms are matched to individual drivers. A highly robust algorithm would perform well for everyone without the need to tune the algorithm to each person or class of people. Applying a z-transform to the raw data is one way to individualize the algorithms. A z-score was calculated for each measure by subtracting the mean and dividing by the standard deviation using the data for each variable, BAC level, and driver combination. Figure 20 shows the effect of this transform for three measures—the standard deviation of lane position for Urban Curves (106), Interstate Curves (205), and Gravel Rural (306) events. The left panels show raw measures and the right panel shows z-transformed variables. The benefit of this form of individualization is revealed by the greater separation of the z-transformed variables across BAC levels.

Figure 21 shows the how well the z-transformed variables separate the experimental conditions—drivers with high BAC tend to cluster in the upper right of the graph with high lane position variability for both events. Surprisingly, the lane position variability in one event is not strongly correlated with that in the other event. The bottom panel shows instances of high and low BAC as blue and red dots, overlaid by a color-coded SVM value function that classifies high and low BAC levels. The right SVM image shows the ADAboost solution addressing those cases that were not classified properly by the first. The z-transformed data from events 106 and 304 effectively identify impairment. This algorithm uses just lane position variability from two events, over total span of only 480 seconds, and classifies BAC level as effectively as the SFST, with accuracy of 80.3%, an AUC of 0.75, and a PPP of 75.0%.

Applying the z-transform to other driving metrics provides a similar benefit to that seen with the standard deviation of lane position. Including z-transformed variables in the algorithms increased their sensitivity for all segments of the drive.

Another means of increasing sensitivity is to add additional variables. Eye data are a particularly promising source of information because such data reflect both executive function and attention management, as well as alcohol-induced drowsiness. Variables addressing gaze concentration, such as percent on road center and horizontal standard deviation of gaze might indicate diminished executive control and Perclos and blink duration, might reflect drowsiness. Eye movement failed to enhance the sensitivity of the algorithms in the freeway and rural segments, but the data suggests potential improvement for the urban segment, although this improvement did not achieve statistical significant, as indicated by overlapping confidence intervals. The perclos and blink duration in Event 103 helped produce an overall accuracy of 81.2 (CI=5.3) and an AUC of 0.84 (CI=0.074).

As a point of comparison, the performance of the decision tree for the urban segment, without cross validation, was very high: accuracy (85.9%), AUC (0.90), and PPP (79.2%). The difference between this performance and that in Table 18 reflects a similar generalization error as seen with the algorithms that were not individualized. Individualized algorithms are no more robust to generalization error than generic algorithms. Compared to the generic algorithm, the individualized algorithms are less robust to the types of roadway. Table 19 shows that sensitivity, measured by AUC, ranges from 0.71 to 0.81; this compares to a much narrower range of 0.63 to 0.65, in Table 18. One explanation for why individualized algorithms are more vulnerable to differences in road types is that the differences between urban and rural segments provide more opportunities for idiosyncratic behavior and driver-specific strategies. The SVM shows similar sensitivity and robustness to the decision tree, both of which performed worst in the freeway segment.

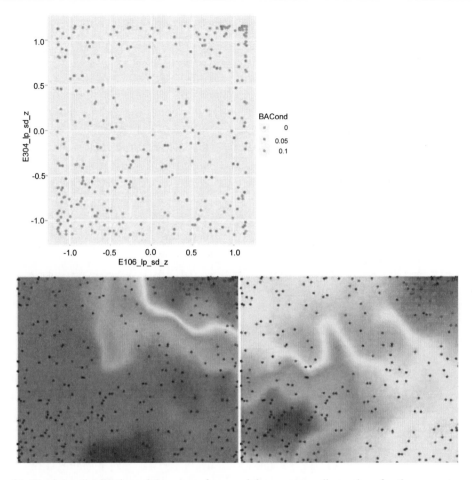

Figure 21. The joint distribution of the z-transform and the corresponding values for the support vector machine classification algorithm.

Table 19. Cross validation to assess robustness defined by generalization error

	Accuracy	AUC	PPP
Urban			
Decision tree	78.6 (5.6)	0.79 (0.10)	68.3 (16.9)
SVM	78.7 (7.4)	0.82 (0.10)	64.5 (25.0)
Freeway			
Decision tree	72.2 (5.2)	0.71 (0.06)	52.2 (25.4)
SVM	71.6 (1.9)	0.68 (0.15)	NA
Rural			
Decision tree	77.6 (3.6)	0.81 (0.048)	68.3 (14.0)
SVM	77.4 (5.4)	0.82 (0.074)	62.8 (11.4)
Field sobriety test cross validation	81.8 (5.9)	0.76 (0.087)	78.4 (15.5)

95% confidence interval in parentheses.

8.4.3. Algorithm robustness, driver characteristics, and alcohol levels

Another approach to robustness considers the degree to which irrelevant variables influence impairment detection. Developing a logistic regression algorithm that detects impairment by combining decision tree output for each segment assessed this possibility. Because this algorithm combines data from all three segments it is very sensitive, detecting impairment with an AUC of .96. With a perfectly robust algorithm, the misclassified instances would not be systematically related to irrelevant factors such as driver age, experience, or weight. Examining the residuals of a logistic regression can assess whether the residuals depend on various conditions. Residuals are the difference between the predicted and actual BAC condition, and the degree various factors affect algorithm performance is an indicator of algorithm robustness.

Of particular interest is the influence of age and driving experience. The first analysis considered age as a continuous variable and no difference in the distribution of residuals across age, $F(1,310)=0.435$, $p=.509$. A similar pattern is seen for weight, $F(1, 310)=1.536$, $p=.216$. A second analysis addressed the possible benefit of greater driving experience associated with age. For this comparison, drivers were divided into two groups: those below 25 years of age and those at or above 25. No statistically significant difference emerged, $F(1, 311)=.14$, $p=.707$.

Another factor that might be expected to confound alcohol detection is drowsiness. Although substantial care was taken to minimize conflicts with drivers' typical sleep schedules, some drivers drove while fatigued. Drowsiness, as measured by the Stanford Sleepiness Scale, was not strongly associated with the residuals when measured before the drive, $F(1, 311)=.955$, $p=.329$, or when measured after the drive, $F(1, 309)=1.457$, $p=.228$. Similar analyses were performed for other potentially confounding factors, such as drive number and scenario number with similar results.

Several variables were strongly related to the residuals: the gaze nystagmus score from the SFST, $F(1, 310)=65.09$, $p<.0001$, and the pre-, $F(1, 311)=121.60$, $p<.0001$, and post-drive, $F(1, 311)=131.22$, $p<.0001$, BAC levels. Figure 22 shows that the algorithm underestimates BAC levels associated with the 0.05% BAC condition and overestimates the level in the 0.10% BAC condition. This simply reflects the criterion used in training the algorithm, where the algorithm differentiated between those above and below 0.08% BAC and so classed those at 0.05% BAC as unimpaired. More interesting is the relationship between the gaze nystagmus and the residuals. This relatively strong relationship suggests that gaze nystagmus is sensitive to indicators of alcohol-related impairment that the algorithm does not capture. This has important practical consequences because combining the vehicle-based estimate of alcohol impairment with gaze nystagmus might produce a much more effective indicator of impairment than either alone. Overall, the analysis of residuals shows that the algorithm is robust to the potentially confounding effects of age, experience, and drowsiness, and is related to BAC levels as expected given the impairment criterion chosen for algorithm development.

8.4.4. Algorithm sensitivity and timeliness

Timeliness refers to how quickly algorithms accumulate sufficient information to judge impairment precisely. Three fundamental considerations govern timeliness: rate of change of impairment, change in sensitivity with accumulation of evidence, and time course of impairment signature. The relationship between the rate of change of the system state relative to the time needed to identify the state is specified by the Nyquist interval, which requires a sampling rate twice that of the bandwidth of the system. Alcohol impairment changes over a time course of hours rather than seconds, so accurate state estimation can be achieved with an algorithm that estimates impairment at a time scale of 20-40 minutes. This contrasts with impairment associated with distraction, which can change much more rapidly and would require an algorithm that detects changes on a time scale of several seconds. Given the long time-constant associated with BAC, this analysis focuses on how much information must be accumulated to provide a sensitive indicator of impairment and the time scale of behavioral signatures needed for real-time interventions.

Sampling theory predicts that measurement uncertainty diminishes with the square root of the number of independent samples, suggesting that algorithm sensitivity will increase with the accumulation of information, but at a diminishing rate. The data from Table 16 plotted in Figure 23 show a general trend toward increasing sensitivity with longer events, but also indicate that longer events provide an increasing benefit. This figure also shows the substantial differences between events, with Urban Drive (102) and Urban Curves (106) being more sensitive than their duration would suggest, contrasting with Interstate Curves (205), which is less sensitive. As noted previously, highly precise impairment detection can occur in eight minutes if the driver encounters situations similar to Urban Curves (106) followed by Dark Rural (304). These results show that timely impairment detection depends on the types of events encountered by the driver, as well as the duration of information accumulation.

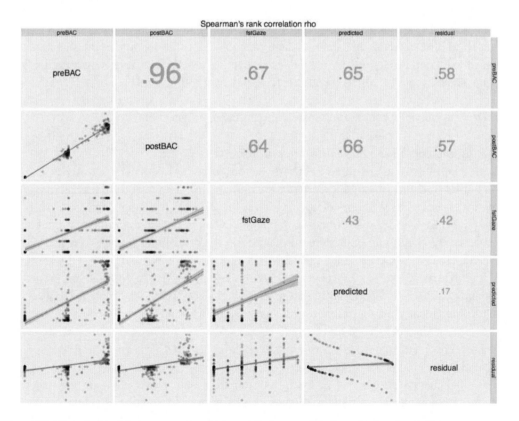

Figure 22. The relationship between algorithm predictions, residuals and related variables.

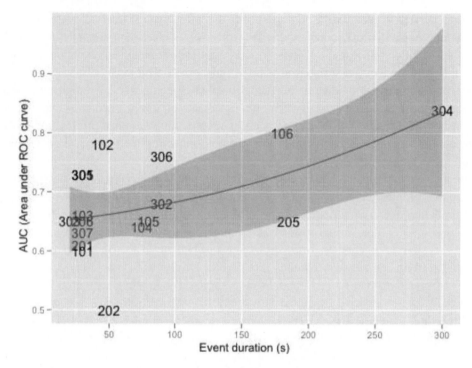

Figure 23. The sensitivity of each event as a function of its duration.

The most sensitive indicators of impairment involve continuous measures cumulated over time, such as the standard deviation of lane position. In addition, important signatures of alcohol impairment are defined by behavior that evolves over a relatively long time horizon, requiring samples of driving behavior that extend over 30 seconds to several minutes—weaving cannot be captured in 10 seconds of data. Even with such constraints, reasonable sensitivity was obtained over approximately three to nine minutes.

8.4.5. Bias and combining algorithms to minimize false alarms

Both robustness and timeliness concern factors affecting algorithm sensitivity and its overall ability to detect impaired drivers *and* avoid labeling unimpaired drivers as impaired. In contrast, bias concerns the tendency to err on the side of detecting impaired drivers *or* avoiding labeling unimpaired drivers as impaired. Generally the algorithms were biased toward correctly identifying unimpaired drivers. PPP, the probability that a driver had a high BAC when labeled as such by the algorithm, ranged between 63% and 68%. This contrasts with approximately 80% for negative predictive performance. Negative predictive performance (NPP) is the probability that a driver had a low BAC when labeled as such by the algorithm. Ideally an algorithm would have high PPP and NPP. One reason for this outcome is the unequal distribution of the data, where 71.5% of the training data were unimpaired cases and the focus was on maximizing the AUC rather than PPP. A cost-sensitive classification process could reverse this tendency.

The fundamental differences between decision trees and SVMs also suggest that they might have complementary strengths. Figure 24 shows how these complementary strengths can be leveraged to minimize false detection of impairment and maximize detection of impaired drivers. The vertical axis represents the probability of detecting an impaired driver, and the horizontal axis represents the chance of a false detection. A prefect algorithm would generate a curve that follows the upper left edge of the plot and an algorithm with no power to discriminate would be a diagonal line from the lower left to the upper right. The area under this curve is AUC. The curves represent the individualized algorithms, and the band around them represents the 95% confidence interval. This band shows that they are largely overlapping, but the blue line, representing the SVM, exceeds the red line of the decision tree when the decision threshold is low. This tendency is also reflected in Table 19, where SVMs produce higher PPP, but the confidence intervals overlap, so it does not represent a statistically significant difference. These data provide tentative evidence that a decision tree might be best when false alarms are of most concern and an SVM when misses are of most concern.

The bias of one algorithm relative to another makes it superior only in the context of the intervention it supports. An algorithm that favors hits over false alarms might be better for certain applications, such as post-drive feedback, but a poor choice for others. Bias describes an important trade-off between algorithms.

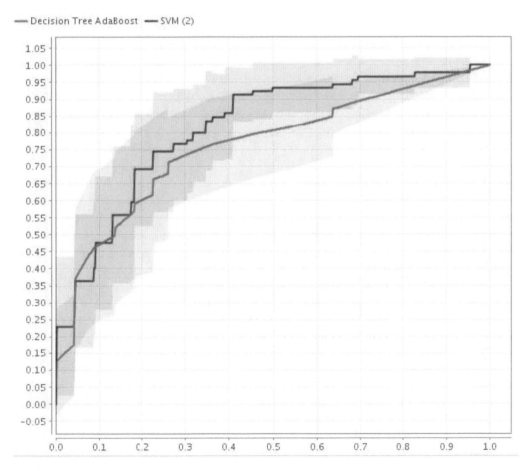

Figure 24. ROC plot for the rural segment showing that the SVM performs better than the decision tree if the decision criterion is low, favoring hits over misses.

8.5. Conclusion

The ultimate aim of this analysis was to assess the feasibility of vehicle-based sensors to detect alcohol-related impairment in real time. The results demonstrate the feasibility of such a system: the sensitivity of an individualized algorithm is comparable to that of the SFST. The behavioral signatures of alcohol impairment that form the basis of this algorithm are consistent with previous studies: diminished lateral and longitudinal control, particularly the standard deviation of lane position. The decision trees fit to individual events reveal a more complex behavioral signature of alcohol: the particular variables depend on the event, the sensitivity depends on the event, and the combination of variables is complex and depends on differences between people and events. Reflecting the complexity of the behavioral signatures, decision tree and SVM algorithms detect alcohol impairment more accurately than a traditional logistic regression using generic measures of average speed, speed variability, and lane position variability.

Analysis of algorithm sensitivity, robustness, and timeliness showed that algorithm sensitivity depends on differences between events, road segments, and drivers. The algorithms are not robust to these factors, which suggests future development must consider how to mitigate and capitalize on these effects. For example, individualizing the algorithms, even with a very simple z-transform of the raw data, substantially increases sensitivity. Similarly, using measures that best reflect alcohol impairment for each event substantially increases sensitivity. Sensitivity depends on the event and its duration. Demanding events tend to be more sensitive, as do longer events. The implication is that the amount of time an algorithm requires to accurately detect impairment depends on the sequence of roadway situations encountered by the driver. The situations included in this study show that an accuracy of 78%, comparable to that of the SFST, can be achieved in approximately eight minutes.

The ultimate aim of impairment-detection algorithms is to support interventions that guide the driver to safer behavior. The desirability and feasibility of any particular algorithm depends on how it meets the particular needs of an intervention. The ideal algorithm would be sensitive, robust, and timely. This study demonstrates that algorithms differ substantially on these dimensions and that design must consider the inevitable tradeoffs. Most importantly, algorithms become more sensitive, but less timely, as measures are integrated over time. The ultimate feasibility of impairment-detection algorithms depends on matching the performance profile of the algorithm to the nature of the intervention.

Conclusion and Implications for Future Research

This study demonstrated that a vehicle-based system using measures of driver behavior can differentiate between drivers with BAC levels above and below 0.08% with a sensitivity similar to the SFST. Because the indicators of alcohol impairment become much stronger at higher levels, the sensitivity would likely increase substantially if the algorithm was used to identify those with BAC levels over 0.150%. These outcomes strongly support the potential of vehicle- based systems to prevent and mitigate alcohol-related crashes.

Table 20 shows the potential of a vehicle-based indicator of impairment, particularly one that is sensitive to high BAC. Lund (2006) used the relative risk curves calculated by Zador, Krawchuk, and Voas (2000) based on the FARS and the 1996 National Roadside Survey (Voas, Wells, Lestina, Williams, & Greene, 1997) and applied those relative risk values to an estimation of the savings in driver fatalities if new vehicle "technologies" succeeded in preventing drivers at certain BACs from driving. Table 20 shows that 66% of alcohol-related fatalities occur with BAC levels above 0.150%, suggesting that the greatest value of a vehicle-based countermeasure lies in detecting high BAC levels, where algorithms are likely to be very sensitive. The exact percent of crashes associated with a given BAC level varies from year to year; therefore, a precise estimate of benefits is difficult to assess. For example, the data from NHTSA suggests a rate of 57% when comparing the number of drivers in fatal crashes with a 0.15 BAC or greater out of all drinking drivers (.01 BAC or greater). This table does not imply that the algorithm in the report would prevent all or even a majority of the crashes associated with high BAC levels, only that most fatalities occur at very high BAC levels where the algorithm is likely to be most sensitive.

The relative contribution of alcohol to degraded performance versus risky behavior has important implications for detecting alcohol impairment with simulator data. People tend to behave better under observation, such as attempting to adhere to speed limits, than they might when they are not being monitored (Evans, 1991; Rowe, et al., 2006). As a consequence, data collected in the simulator may only capture the effects of alcohol on motivated performance and not the effects on behavior or typical driving performance. The result is that simulator data may not capture the full range of responses that an algorithm could use to identify alcohol impairment. Given that simulator data may reflect alcohol-related changes in performance and not behavior, there may be sensitive indicators of alcohol impairment, such as speeding, that might not be revealed in the simulator. Naturalistic studies of driving, such as the Strategic Highway Research Program 2 (SHRP2), could complement the data collected in driving simulators and provide a valuable platform for assessing impairment-detection algorithms.

On the basis of this research, standard deviation of lane position and average speed were shown to be reliable measures of impairment that can be feasibly captured over a number of driving situations, and appear robust enough to be useful in future vehicle-based countermeasures. Minimum speed, as well as standard deviation of lane position and speed, are useful indicators that might have particular utility in alcohol warning monitors designed to provide feedback to drivers.

Table 20. Potential lives saved in 2004 if driver BACs had been limited to <0.08% [adapted from Lund (2006)].

Driver %BAC	Fatalities	Estimated reduction
.150+	8,629	6,540
.100 – .149	3,430	1,143
.080 – .099	1,083	203
ALL	13,142	7,886

The results of this study have implications for future research. One general finding is that data mining techniques, such as decision trees and support vector machines, can extract information from driving performance data that might otherwise be lost. Such information supports more timely and sensitive algorithms. Further exploration of such techniques with different types of impairment has great promise.

A second general finding is that the driving context strongly influences impairment-detection performance. Contrary to many previous simulator studies of alcohol-impaired driving, this study used a representative series of 19 events over three types of roadway situations. These events revealed that impairment detection depends on the type of event. Because driving is a satisfying rather than optimizing activity, drivers can take many paths through low-demand situations that are all satisfactory. This variety of satisfactory responses masks impairment. The variety of events also requires a greater variety of measures to capture the relevant behavior in each event. All of these findings imply that detecting alcohol-related impairment, and impairment detection more generally, depends on the driving situation. Algorithm development needs to consider roadway situations as much as it needs to consider the drivers' perceptual, motor, and decision-making response to the impairment.

A third general finding is that individual differences strongly influence driving performance indicators and the sensitivity of the algorithm. The analysis of the algorithms confirmed that it is possible to create a diagnostic algorithm that is not tailored to an individual driver. Under the conditions tested, that generalized algorithm was timely, robust, and nearly as accurate as the individualized algorithm. However, tailoring algorithms to individuals has the potential to substantially enhance algorithm sensitivity. An important consideration regarding the degree to which algorithms should consider individual differences concerns the role of algorithms in promoting safer driving. If the ultimate aim of vehicle-based systems is to identify and mitigate dangerous driving behavior, tailoring the algorithm to the individual might lead the algorithm to neglect drivers with consistently unsafe behavior. This suggests tailoring algorithms to individuals' needs to differentiate between satisfactory, but different, responses and unsafe behavior. In general, this study demonstrates that vehicle-based algorithms to detect impairment are feasible, but that their performance depends on a careful tailoring of the algorithm to the drivers and roadway situation. The success of future impairment-detection algorithms will likely depend on understanding how the impairment interacts with different types of drivers, trips, and roadway situations. An important research question concerns identifying classes of roadway situations and classes of drivers so that the algorithms do not need to be tailored to each individual driver and each individual roadway situation. Eye movement data might be particularly useful to consider in this context.

These results support the long-term research objective of using algorithms that detect impairment to provide drivers with feedback that will discourage or prevent drinking and driving. Ultimately the distraction-detection algorithms developed in this study could support a range of vehicle- based interventions to prevent alcohol-related crashes. Such interventions could include limiting drivers' ability to drive dangerously (e.g., lockout distractions or limit speed), providing feedback to impaired drivers that may motivate them to pull over or drive more cautiously, adjusting crash warning systems to provide an earlier warnings, or providing long-term feedback that highlights dangerous driving. Such interventions all depend on a reliable means of detecting alcohol impairment using driver behavior data, which this study demonstrates as being feasible.

The promising results associated with alcohol-related impairment detection suggest other types of impairment detection might also hold promise, most notably distraction and drowsiness. As such algorithms and associated interventions are developed, their joint performance must be considered. One such consideration is the definition of a false alarm. False alarms and misses need to be interpreted in the context of how the impairment detection will be used. If the ultimate goal of a system is to identify impaired behavior rather than inferring the presence of alcohol, then the meaning of false alarms and misses changes substantially. An alcohol-impairment algorithm might be sensitive to drowsiness and so could provide valuable information even if the impairment it detects is not the one it was originally designed to detect. The ultimate definition of algorithm sensitivity might need to depend on the intervention it supports, just as the required sensitivity depends on the intervention.

ACRONYMS AND ABBREVIATIONS

ANOVA	analysis of variance
BAC	blood alcohol concentration
CI	confidence interval
EEG	electroencephalogram
FSA	force sensor arrays
NADS	National Advanced Driving Simulator
NHTSA	National Highway Traffic Safety Administration
OEM	original equipment manufacturer
PCA	principal components analysis
PPP	positive predictive performance
QFV	Quantity-Frequency-Variability scale
ROC	receiver operating characteristic
RPM	rotations per minute
SD	standard deviation
SDLP	standard deviation of lane position
SE	steering entropy
SSS	Stanford Sleepiness Scale
SFST	standardized field sobriety test
SUV	sport-utility vehicle
SVD	singular value decomposition
TH	time headway

ACKNOWLEDGMENTS

The project benefited from the contributions of Mark Gilmer and Tim Brewer from Intoximeter, who arranged for a loan of the Alco-Sensors for this study.

REFERENCES

Amari, S. & Wu, S. (1999). Improving support vector machine classifiers by modifying kernel functions. *Neural Networks*, *12*(6), 783-789.

Arnedt, J. T., Wilde, G. J. S., Munt, P. W. & MacLean, A. W. (2001). How do prolonged wakefulness and alcohol compare in the decrements they produce on a simulated driving task? *Accident Analysis and Prevention*, *33*(3), 337-344.

Babor, T. F., de la Fuente, J. R., Saunders, J. & Grant, M. (1992). *AUDIT: The alcohol use disorders identification test: Guidelines for use in primary health care.* Geneva, Switzerland: World Health Organization.

Bergasa, L. M., Nuevo, J., Sotelo, M. A., Barea, R. & Lopez, M. E. (2006). Real-time system for monitoring driver vigilance. *IEEE Transactions on Intelligent Transportation Systems*, *7*(1), 63-77.

Blincoe, L., Seay, A., Zaloshnja, E., Miller, T., Romano, E., Luchter, S., et al. (2002). *The Economic Impact of Motor Vehicle Crashes, 2000* (No. DOT HS 809 446). Washington, DC: National Highway Traffic Safety Administration.

Blomberg, R. D., Peck, R. C., Moskowitz, H., Burns, M. & Fiorentino, D. (2005). *Crash risk of alcohol invovlved driving: A case control study.* Stamford, CT: Dunlap and Associates.

Boer, E. R. (2001). Behavioral entropy as a measure of driving performance. In D. V. McGehee, J. D. Lee & M. Rizzo (Eds.), *Proceedings of the International Driving Symposium on Human Factors in Driving Assessment, Training, and Vehicle Design* (225-229). Public Policy Center, Iowa City, IA.

Brookhuis, K. A. & De Waard, D. (1993). The use of psychophysiology to assess driver status. *Ergonomics*, *36*(9), 1099-1110.

Brookhuis, K. A., De Waard, D. & Fairclough, S. H. (2003). Criteria for driver impairment. *Ergonomics*, *46*(5), 433-445.

Burns, M. & Dioquino, T. (1997). *A Florida Validation Study of the Standardized Field Sobriety Test (SFST) Battery* (No. Florida Department of Transportation Rep.No. AL-97-05-14-01). Los Angeles, CA: Southern California Research Institute.

Burns, M. & Moskowitz, H. (1977). *Psychophysical Tests for DWI Arrest* (No. US Department of Transportation Rep. No. DOT HS-802 424). Washington, DC: National Highway Traffic Safety Administration.

Byun, H. & Lee, S. W. (2002). Applications of Support Vector Machines for Pattern Recognition: A Survey. *Pattern Recognition with Support Vector Machines, First International Workshop, SVM 2000*, 213-236.

Chung, T., Colby, S. M., Barnett, N. P. & Monti, P. M. (2002). Alcohol use disorders identification test: Factor structure in an adolescent emergency department sample. *Alcoholism: Clinical and Experimental Research*, *26*(2), 223-231.

Conley, T. B. (2001). Construct validity of the MAST and AUDIT with multiple offender drunk drivers. *Journal of Substance Abuse Treatment*, *20*(4), 287-295.

de Waard, D., Brookhuis, K. A. & Hernandez-Gress, N. (2001). The feasibility of detecting phone-use related driver distraction. *International Journal of Vehicle Design*, *26*(1), 85- 95.

Efron, B. & Gong, G. (1983). A leisurely look at the bootstrap, the jackknife, and cross-validation. *American Statistician*, *37*(1), 36-48.

Evans, L. (1991). *Traffic Safety and the Driver*. New York: Van Nostrand Reinhold.

Fairclough, S. H. & Graham, R. (1999). Impairment of driving performance caused by sleep deprivation or alcohol: A comparative study. *Human Factors*, *41*(1), 118-128.

Fell, J. C., Tippetts, A. S. & Voas, R. B. (2009). Fatal traffic crashes involving drinking drivers: what have we learned? *Annals of Advances in Automotive Medicine*, *53*, 63-76.

Feng, C. X. J., Yu, Z. G. S., Emanuel, J. T., Li, P. G., Shao, X. Y. & Wang, Z. H. (2008). Threefold versus fivefold cross-validation and individual versus average data in predictive regression modelling of machining experimental data. *International Journal of Computer Integrated Manufacturing*, *21*(6), 702-714.

Grace, R. & Suski, V. (2001). Improving safety for drivers and fleets: Historical and innovative approaches. In D. V. McGehee, J. D. Lee & M. Rizzo (Eds.), *Proceedings of the International Driving Symposium on Human Factors in Driving Assessment, Training, and Vehicle Design* (pp. 345-350). Iowa City: IA: Public Policy Center.

Harris, D. H. (1980). Visual detection of driving while intoxicated. *Human Factors*, *22*(6), 725-732.

Holloway, F. A. (1994). *Low-dose Alcohol Effects on Human Behavior and Performance: A Review of Post-1985 Research*. Oklahoma City, OK: Civil Aeromedical Institute, Federal Aviation Administration.

Kennedy, R. S., Turnage, J. J., Rugotzke, G. G. & Dunlap, W. P. (1994). Indexing Cognitive Tests to Alcohol Dosage and Comparison to Standardized Field Sobriety Tests. *Journal of Studies on Alcohol*, *55*(5), 615-628.

Kennedy, R. S., Turnage, J. J., Wilkes, R. L. & Dunlap, W. P. (1993). Effects of Graded Dosages of Alcohol on 9 Computerized Repeated-Measures Tests. *Ergonomics*, *36*(10), 1195-1222.

Lacey, J. H., Kelley-Baker, T., Furr-Holden, D., Voas, R. B., Romano, E., Torres, P., et al. (2009). *2007 National Roadside Survey of Alcohol and Drug Use by Drivers: Alcohol Results*. Washington, DC: National Highway Traffic Safety Administration.

Lenne, M. G., Dietze, P., Rumbold, G. R., Redman, J. R. & Triggs, T. J. (2003). The effects of the opioid pharmacotherapies methadone, LAAM and buprenorphine, alone and in combination with alcohol, on simulated driving. *Drug and Alcohol Dependence*, *72*(3), 271-278.

Liang, Y., Reyes, M. L. & Lee, J. D. (2007a). Non-intrusive detection of driver cognitive distraction in real-time using Bayesian networks. *Transportation Research Board*(2018), 1-8.

Liang, Y., Reyes, M. L. & Lee, J. D. (2007b). Real-time detection of driver cognitive distraction using Support Vector Machines. *IEEE Intelligent Transportation Systems*, *8*(2), 340-350.

Lim, T. S., Loh, W. Y. & Shih, Y. S. (2000). A comparison of prediction accuracy, complexity, and training time of thirty-three old and new classification algorithms. *Machine Learning*, *40*(3), 203-228.

Linnoila, M. & Mattila, M. J. (1973). Drug interaction on driving skills as evaluated by laboratory tests and by a driving simulator. *Pharmakospsychiatrie Neuro-Psychopharmakologie*, *6*, 127-132.

Lund, A. K. (2006). Eliminating alcohol-impaired driving: Potential effects of technology applied to the general population, Insurance Institute for Highway Safety *International Technology Symposium*. Albuquerque, NM.

Lund, A. K. & Wolfe, A. C. (1991). Changes in the incidence of alcohol-impaired driving in the United States, 1973-1986. *Journal of Studies on Alcohol*, *52*(4), 293-301.

Marple-Horvat, D. E., Cooper, H. L., Gilbey, S. L., Watson, J. C., Mehta, N., Kaur-Mann, D., et al. (2008). Alcohol badly affects eye movements linked to steering, providing for automatic in-car detection of drink driving. *Neuropsychopharmacology*, *33*, 849 – 858.

Moskowitz, H., Ziedman, K. & Sharma, S. (1976). Visual search behavior while viewing driving scenes under the influence of alcohol and marihuana. *Human Factors*, *18*(5), 417-431.

Nakayama, O., Futami, T., Nakamura, T. & Boer, E. R. (1999). SAE Technical Paper Series: Development of a steering entropy method for evaluating driver workload. *Human Factors in Audio Interior Systems, Driving, and Vehicle Seating, SP-1426*.

National Highway Traffic Safety Administration. (2009a). Fatality Analysis Reporting System (FARS). Retrieved March 30, 2009, from ftp://ftp.nhtsa.dot.gov/fars/

National Highway Traffic Safety Administration. (2009b). Fatality Analysis Reporting System (FARS). Retrieved June, 2009, from ftp://ftp.nhtsa.dot.gov/fars/Ogden, E. J. D. & Moskowitz, H. (2004). Effects of alcohol and other drugs on driver performance. *Traffic injury Prevention*, *5*, 185-198.

Peters, B. & van Winsum, W. (1998). *SAVE: System for effective Assessment of the driver state and Vehicle control in Emergency situations*: R&D Programme Telematics: EU.

Pollard, J. K., Nadler, E. D. & Stearns, M. D. (2007). *Review of Technology to Prevent Alcohol- Impaired Crashes (TOPIC)* (No. DOT HS 810 827). Washington, DC: US Department of Transportation, National Highway Traffic Safety Administration.

Quinlan, J. R. (1996). Improved use of continuous attributes in C4.5. *Journal of Artificial Intelligence Research*, *4*, 77-90.

Rakauskas, M. E., Ward, N. J., Boer, E., Bernat, E. M., Cadwallader, M. & Patrick, C. J. (2008).

Combined effects of alcohol and distraction on driving performance. *Accident Analysis & Prevention*, *40*, 1742–1749.

Roehrs, T., Beare, D., Zorick, F. & Roth, T. (1994). Sleepiness and ethanol effects on simulated driving. *Alcoholism-Clinical and Experimental Research*, *18*(1), 154-158.

Rowe, S. Y., Olewe, M. A., Kleinbaum, D. G., McGowan JR, J. E., McFarland, D. A., Rochat, R., et al. (2006). The influence of observation and setting on community health workers' practices. *International Journal for Quality in Health Care*, *18*(4), 299–305.

Rubenzer, S. J. & Stevenson, S. B. (2010). Horizontal Gaze Nystagmus: A Review of Vision Science and Application Issues. *Journal of Forensic Science*, *55*(2).

Saad, L. (2003). Moderate drinking on the rise: The Gallup Organization, Gallup News Service, Poll Analyses.

Stuster, J. W. (1997). *The Detection of DWI at BACs Below 0.10*. Washington, DC: US Department of Transportation, National Highway Traffic Safety Administration.

Stuster, J. W. (1999). *Development of an Automated DWI Detection System*. Washington, DC: NHTSA.

Stuster, J. W. (2006). Validation of the standardized field sobriety test battery at 0.08% blood alcohol concentration. *Human Factors*, *48*, 608-614.

Tippetts, A. S. & Voas, R. B. (2002). Odds that an involved driver was drinking: Best indicator of an alcohol-related crash? In D. R. Mayhew & C. Dussault (Eds.),

Proceedings of the 16th International Conference on Alcohol, Drugs and Traffic Safety (Vol. 1, 67-71). Montreal, Canada: Société de l'assurance automobile du Québec.

Tzambazis, K. & Stough, C. (2000). Alcohol impairs speed of information processing and simple and choice reaction time and differentially impairs higher-order cognitive abilities. *Alcohol and Alcoholism, 35*(2), 197-201.

Vanakoski, J., Mattila, M. J. & Seppala, T. (2000). Driving under light and dark conditions: effects of alcohol and diazepam in young and older subjects. *European Journal of Clinical Pharmacology, 56*(6-7), 453-458.

Vapnik, V. N. (1995). *The nature of statistical learning theory*. New York: Springer. Verwey, W. B. & Veltman, H. A. (1996). Detecting short periods of elevated workload: A comparison of nine workload assessment techniques. *Journal of Experimental Psychology-Applied, 2*(3), 270-285.

Voas, R. B., Wells, J., Lestina, D., Williams, A. & Greene, M. (1997). Drinking and driving in the US: The 1996 National Roadside Survey. In C. Mercier-Guyon (Ed.), *Proceedings of the 14th International Conference on Alcohol, Drugs and Traffic Safety - T97, Annecy, 21- 26 September 1997* (Vol. *3*, 1159-1166). Annecy, France: Centre d'Etudes et deviation Recherches en Médecine du Trafic.

Voas, R. B., Wells, J., Lestina, D., Williams, A. & Greene, M. (1998). Drinking and driving in the United States: The 1996 National Roadside Survey. *Accident Analysis and Prevention, 30*(2), 267-275.

Ward, N. J. (2006). *Alcohol, Traffic Safety, and the Conception of Vehicle-based Systems to Detect Driver Impairment*.

Wolfe, A. C. (1974). *1973 US national roadside breath testing survey: procedures and results*. Ann Arbor, MI: University of Michigan Safety Research Institute.

Zador, P. L., Krawchuk, S. A. & Voas, R. B. (2000). Alcohol-related relative risk of driver fatalities and driver involvement in fatal crashes in relation to driver age and gender: An update using 1996 data. *Journal of Studies on Alcohol, 61*(3), 387-395.

Zador, P. L., Krawchuk, S. A. & Voas, R. B. (2001). Alcohol-related relative risk of driver fatalities and driver involvement in fatal crashes in relation to driver age and gender: An update using 1996 data. *Journal of Studies on Alcohol,* (May), 387-395.

End Notes

[1] Visual occlusion is a research technique that blocks the driver's ability to view the driving environment except when the driver presses a button. On pressing the button, the driving environment is revealed for one second.

[2] For the purposes of this project, we considered relative change in speed associated with driving too fast only as an increase of 20%.

[3] A total of nine participants passed the screening but were not scheduled for a second visit.

[4] Note that this was an a-priori adjustment to the alpha level to control for type I error rather than a particular statistical approach for control. Although arbitrary, we believe it is reasonable.

INDEX

D

E

S

T

U